Brandeis meets Gutenberg Vol. 2

Res Publica

Öffentliches und internationales Recht

Herausgegeben von Udo Fink, Dieter Dörr und Rolf Schwartmann

Band 22

PETER LANG

Udo Fink / Russell L. Weaver /
Ines Gillich (eds.)

# Brandeis meets Gutenberg Vol. 2

## Contemporary Threats to Free Speech

PETER LANG

**Bibliographic Information published by the Deutsche Nationalbibliothek**
The Deutsche Nationalbibliothek lists this publication in the Deutsche Nationalbibliografie; detailed bibliographic data is available online at http://dnb.d-nb.de.

**Library of Congress Cataloging-in-Publication Data**
A CIP catalog record for this book has been applied for at the Library of Congress.

ISSN 1614-838X
ISBN 978-3-631-87094-5 (Print)
E-ISBN 978-3-631-87117-1 (E-PDF)
E-ISBN 978-3-631-87118-8 (EPUB)
DOI 10.3726/b19319



Peter Lang – Berlin · Bern · Bruxelles · New York · Oxford · Warszawa · Wien



This publication has been peer reviewed.

www.peterlang.com

# Table of Contents

# List of Contributors

**Morgan Cloud**
Charles Howard Candler Professor of Law, Emory University.

**Udo Fink**
Professor of Public Law, European and Public International and International Business Law, Johannes Gutenberg University of Mainz.

**Ines Gillich**
Professor of Public Law, European Law and Public International Law, University of Rostock.

**András Koltay**
Professor of law, University of Public Service (Budapest) and Pázmány Péter Catholic University (Budapest).

**Michael Losavio**
Associate Professor, Department of Criminal Justice and instructor, Department of Computer Science and Engineering, University of Louisville.

**Iris Nguyên Duy**
Professor at the Department of Law, School of Business and Law, University of Agder, Norway

**Russell L. Weaver**
Professor of Law & Distinguished University Scholar, University of Louisville, Louis D. Brandeis School of Law. Professor Weaver wishes to thank the University of Louisville's Distinguished Scholar program for supporting his research activities.

# Vorwort: Das "Free Speech Discussion Forum" (Mainz, 2021)

"Brandeis meets Gutenberg" – Unter diesem Motto steht der langjährige fruchtbare wissenschaftliche Austausch zwischen der Louis D. Brandeis Law School der University of Louisville, Kentucky und der rechtswissenschaftlichen Abteilung der Johannes Gutenberg-Universität Mainz. Der Austausch ist auf den unterschiedlichsten Ebenen angesiedelt. Eine Konstante ist der jährlich stattfindende Austausch von Dozenten. Kollegen aus Louisville lehren in Mainz US-amerikanisches Recht und Mainzer Kollegen bereichern das Lehrangebot der Partneruniversität um Vorlesungen im deutschen und im europäischen Recht. Auch das Völkerrecht wird sowohl aus der US-amerikanischen wie aus der deutschen und europäischen Perspektive beleuchtet. Daneben finden Vorträge zu den verschiedensten Themen statt, die entweder vor Studenten oder als „faculty speeches" gehalten werden. Durch den persönlichen Kontakt der Herausgeber dieses Bandes ist Mainz in ein breites Netzwerk US-amerikanischer Hochschulen eingebunden, das jedes Jahr in einer weltweit von Kollegen besuchten Konferenz der South Eastern American Law Schools gipfelt. Im Gegenzug finden in Mainz und anderen Orten in Europa vergleichbare Konferenzen mit Kollegen aus der ganzen Welt statt.

All diese Programme leisten einen wichtigen Beitrag zum internationalen Profil der beteiligten Universitäten. Verantwortlich für den wissenschaftlichen Austausch sind Professor Russell L. Weaver, Professor of Law and Distinguished University Scholar an der University of Louisville, und Professor Dr. Udo Fink, Inhaber des Lehrstuhls für Öffentliches Recht, Völker- und Europarecht sowie Internationales Wirtschaftsrecht und Louisville-Beauftragter der rechtswissenschaftlichen Abteilung an der Universität Mainz.

Im Rahmen dieser Kooperation fand im Juni 2021 ein "Free Speech Discussion Forum" an der Universität Mainz als zweitägiges digitales Forum statt. Namhafte Rechtswissenschaftler aus den USA, Deutschland, Luxemburg, Ungarn, Norwegen und England referierten und diskutierten über "Gegenwärtige Herausforderungen der Meinungsfreiheit." Free speech und verwandte Themen im Zusammenhang mit der Meinungsfreiheit und den Medien sind schon mehrfach Gegenstand solcher Foren gewesen. Dies hängt mit den wissenschaftlichen Neigungen der Programmbeauftragten und den daraus resultierenden Kontakten mit in- und ausländischen Kollegen, aber auch mit der außerordentlichen Brisanz der Materie insgesamt zusammen, die immer wieder spannenden Stoff für den akademischen Austausch bietet.

Der vorliegende Tagungsband fasst die wichtigsten Ergebnisse der Konferenz zusammen und knüpft damit auch an den ersten Band "Brandeis meets Gutenberg", der 2012 erschienen ist, an.

*Udo Fink Ines Gillich*

Russell L. Weaver

# Introduction: Brandeis Meets Gutenberg (Vol. II)

The Free Speech Discussion Forum is an annual event that is designed to bring together prominent free scholars from all over the world to discuss cutting-edge free speech issues. The 2021 meeting was scheduled to be held in Mainz,Germany, because of its connection to Johannes Gutenberg, the inventor of the printing press. Unfortunately, because of the Covid-19 pandemic, the participants were unable to meet in person at Johannes Gutenberg University, and were forced to move their discussions online.

In keeping with the thought-provoking location, two major topics were discussed at the 2021 forum: "Robotic Speech" and "Contemporary Threats to Speech." The "Robotic Speech" topic included examination of such things as "chatbots," data driven speech, "deep fakes," and the potential for internet speech to interfere with democratic elections. The "Contemporary Threats to Speech" focused on a variety of topics, including the growth and power of social media platforms.

Included in this compendium of articles is Professor Dr. Udo Fink's article, *Robotic Speech as Fake News?* Professor Fink notes so-called "fake news" has become a hallmark of modern communication because of the internet, and he notes how organizations like Cambridge Analytica have used bots to create fake online identities, as well as to generate and disseminate content to other users of social platforms, and interact with others. In addition, he expresses concern that it can be difficult to distinguish between human speech and bot speech, and he is not convinced that the fundamental right to engage in journalism should apply to bots, and he doubts that they can form and express opinions like human beings. Ultimately, he concludes that they should be subject to regulation. At the very least, "regulation for robots should guarantee their transparency and neutrality regarding the content produced."

Dr. Ines Gillich's article, entitled *"Algorithmic Journalism" and Freedom of the Press*, distinguishes "algorithmic journalism," which "uses AI-driven tools to generate coherent and fluent news content," from other forms of journalism who uses "assisting software for journalistic activities." She defines "algorithmic journalism" as programs that "carry out autonomous tasks by researching, editing, publishing and/or distributing information according to specified variables to specific recipients without human intervention in the process after the final programming or editing." Basically, the algorithms "automatically extract figures from huge data silos," and "automatically convert them into readable stories." She argues that the fundamental right to free expression "applies only to human communication," and therefore "it is questionable whether its ambit also extends to communication generated by an algorithm based on programmed parameters." In regard to the protections for journalism, she views that right as embracing both "rights and responsibilities," and she doubts that algorithms can satisfy the "journalistic due diligence requirements applicable for anthropogenic journalism." As a result, she argues that algorithmic journalism can be subjected to restrictions.

My article, *Robotic Speech and the First Amendment*, also deals with the problem of robotic speech, but views the issues differently than Professor Fink. On first thought, my tendency was to agree with Professor Fink. After all, free speech is important to the democratic process, and is the method by which individuals engage in the democratic process. Since robots are not people, why should we protect their speech? However, as I thought about the problem more deeply, I began to see the issues differently, and it became more difficult to separate robotic speech from human speech. Indeed, some robotic speech is nothing more than a megaphone. In other words, a single individual might use robots to widely disseminate his thoughts and ideas. The rich and powerful have historically controlled the media (not only print, but radio and television), and have therefore had the ability to mass disseminate their ideas. Most people never had that ability or luxury. So, to some extent, for ordinary individuals, the ability to distribute their ideas mechanically is an equalizer. Moreover, when bots create new content, they almost never function in a completely autonomous fashion. Rather, their abilities are limited and controlled by their programmers. So, it strikes me that the problem of how to handle robotic speech is neither simple nor easy.

Professor András Koltay's contribution is entitled *The Notion of the 'Press' with Regard to Freedom of the Press*. He notes the difficulties that arise when scholars try to define "the press," but contends that this definition is essential. Society needs to distinguish between "the press" and other speakers if the press is entitled to special rights, and correspondingly is subject to particular duties and obligations that apply to the press, but do not necessarily apply to other speakers. He argues that the press should be defined by its "functions and tasks" rather than by the technology it uses. Among those "functions and tasks" are the goal of "providing the public with information on public issues." But he also argues that the definition of the "press" necessarily includes other things, including proof of a "professional operation" and the "recognition and observance of professional norms."

Professor Michael Losavio's *Freedom of Speech, Inviolability and Democratic Cyber Security – an End or the Beginning* examines a different aspect of the technology problem. Although the "electronic frontier has brought us to new opportunities for the exchange of ideas, free expression and free association," it also "offers opportunities for suppression, abuse and criminality, especially through attacks by people against the ideas of those with whom they disagree." This is due to aspects of ICT that may injure anonymity, privacy of association, misattribution, mis-inference and open pathways for attack from afar. Whether breach of privacy comes from erroneous inference, or from hacked communications resulting from artificial intelligence/machine learning systems, the injury for speaking is present and evident. Cyber security is essential to protecting rights, yet it has not been subject to either full democratic accountability nor distributed responsibility. One proposed solution seems to run counter to the nation's free speech tradition through the creation of "a national firewall – a series of controlled access points for all Internet traffic in and out of a country – with defined transit for all electronic communications using that network. Technology now exists for the blockage of certain types or origins of communications, and, increasingly, the analysis of those communications as to protocol, content and metadata, especially social network analysis. It can serve to protect the country from malicious use of the system, such as from hostile code and malware." But the risk is its application to ideas and associations deemed harmful to the state. This technical monitoring solution may negate the promise of the Internet to bring us together over new ideas. The Internet and computing are now so ubiquitous that they still offer great opportunities for the exchange of ideas, even under state and private attacks. A better solution is democratic, distributed technologies of cyber security for use by people. These can protect those opportunities, and those brave enough to speak on unpopular matters as to, perhaps, build a better world.

Professor Morgan Cloud's article, *Analog Rights for Expressive Digital Property*, proposes a new model for applying constitutional protections to some expressive digital property. When expressive property is protected by the First Amendment's right to receive information, both the First and Fourth Amendments impose limits to government authority to search for and to use that information as evidence in criminal prosecutions. Fourth Amendment rules offer only limited protections for protected expressive property, however, in part because searches of digital devices often must exceed the scope authorized by judicial search warrants. As a result, government searchers may find information that might be used as evidence against the suspect, but that is protected by the fundamental First Amendment right to receive ideas. A document supporting violent sedition, for example, cannot be criminalized unless it is "directed to inciting or producing imminent lawless action and is likely to incite or produce such action." To the extent that this constitutional privilege prohibits criminalization of possession of this document, it also should generate a bright line First Amendment exclusionary rule that eliminates the ambiguities existing under current Fourth Amendment rules. First Amendment substantive rights exist and can provide constraints on government actors independent of contemporary Fourth Amendment procedural requirements.

Professor Iris Nguyên Duy's article, *Contemporary Threats to Freedom of Expression in Norway in the Context of the 2021 Parliamentary Election and the COVID-19 Pandemic*, takes us in a different direction. In that article, she focuses on a number of hot-button issues in Norway, including the digital manipulation of electors by national political actors, digital threats by foreign actors, as well as the fight against online hate speech, and problems created by the pandemic during the 2021 parliamentary elections. She then discusses Norway's approach to these problems, and notes that the Norwegian response to these problems has been rather "low-key" (but more intense in the case of hate speech) in order to avoid a chilling effect on speech. However, she believes that reforms to the electoral process are on their way, and she expects Norway to incorporate the coming EU-legislation on the regulation of digital platforms. As part of their review, Norway will consider measures designed to promote the broadest possible participation in the public debate, but ones that will hamper the spread of fake news and hate speech.

Udo Fink, Mainz

# Robotic Speech as Fake News?

## I. Introduction

The phenomenon of fake news is not new, but it became a trademark of modern communication through the Internet. Former U.S. president Donald Trump accused newspapers, TV Channels and even elections officials to disseminate fake news.[1] Activities of the now defunct British consulting company Cambridge Analytica and similar organisations in Russia, China and other countries put the focus of this debate on social media.

According to the testimony of a former Cambridge Analytica employee given to the European Parliament in an official hearing, Cambridge Analytica, sought to heavily influence public opinion during the Brexit campaign[2] and was also involved in the 2016 U.S. presidential elections.[3] We have no proper results yet, but the 2020 U.S. elections were obviously also seriously affected by targeted opinion-making through social bots.[4]

Companies like Cambridge Analytica use social media platforms like Facebook and Twitter.[5] Giving the apps installed on these platforms permission

---

1 For example, on May 9, 2018 President Trump mentioned on twitter: "The Fake News is working overtime. Just reported that, despite the tremendous success we are having with the economy & all things else, 91 % of the Network News about me is negative (Fake). Why do we work so hard in working with the media when it is corrupt? Take away credentials?".

2 In his testimony to the European Parliament's Committee on Civil Liberties, Justice and Home Affairs on 4 June 2018, Christopher Wylie said: 'I don't believe Brexit would have happened were it not for the data targeting technology and network of actors set up by Cambridge Analytica.'

3 Researchers estimate, that in 2016, nearly 19 % of all tweets related to the U.S. elections were generated by bots, see Bessi, A. & Ferrara, E. First Monday, https://doi.org/10.5210/fm.v21i11.7090 (2016).

4 See Pascal Podvin, Election 2020: The devastating social impact of bad bots, Security Magazine, November 2nd, 2020, https://www.securitymagazine.com/articles/93810-election-2020-the-devastating-social-impact-of-bad-bots.

5 Kevin Granville, 'Facebook and Cambridge Analytica: What You Need to Know as Fallout Widens?', *The New York Times* (19 March 2018) <https://www.nytimes.com/2018/03/19/technology/facebook-cambridge-analytica-explained.html>.

to access information of the users and their friends leads to the revelation of the personal data of millions of people, making the bots capable of using the data profiles generated to send targeted information or disinformation in order to try and influence voters or human opinion makers.

One of the major tools companies like Cambridge Analytica use are bots. These are software applications that run automated tasks over the Internet. Bots can be used to create fake online identities. For example, they can be used to operate social media accounts which automatically generate or disseminate content to other users of social platforms or otherwise interact with their accounts.[6] Social bots are particularly active on Twitter, but are also found on many other social media platforms that increasingly form part of the system of political communication in many countries. Some scholars have argued that 66 % of all tweets on popular news and current event websites are made by suspected bots. A Russian so-called 'troll farm' with close ties to the Russian Government, the St. Petersburg-based 'Internet Research Agency', has become notorious for making widespread use of bots in its attempts to manipulate public opinion. It employs dozens of specialists to spread its preferred views on Facebook, YouTube and elsewhere.[7] In China, the so-called '50 Cent Army' uses a massive number of personal accounts to flood the Internet with comments.[8]

## II. Brief History of Fake News

Fake news campaigns are not necessarily linked with modern communication technology. History is full of examples fake news campaigns, only ways and means changed. Was it first neighboring gossip and false testimony of political acting persons, the invention of the printing press by Johannes Gutenberg in 1493 amplified the dissemination of disinformation. In 1835, for example, the New York based newspaper "The Sun" published several articles about the

---

6 Samuel Woolley and Marina Gorbis, 'Social media bots threaten democracy. But we are not helpless', *The Guardian* (17 October 2017) <https://www.theguardian.com/commentisfree/2017/oct/16/bots-social-media-threaten-democracy-technology>.

7 Dan Mangan and Mike Calia, 'Special counsel Mueller: Russians conducted 'information warfare' against US during election to help Donald Trump win', *CNBC* (16 February 2018) <https://www.cnbc.com/2018/02/16/russians-indicted-in-special-counsel-robert-muellers-probe.html>.

8 Michael Bristow, 'China's internet "spin doctors"', *BBC News* (16 December 2008) <http://news.bbc.co.uk/2/hi/asia-pacific/7783640.stm>.

discovery of life on the moon, complete with illustrations of humanoid bat-creatures and bearded blue unicorns presumably discovered by the astronomer Sir John Hershel.[9]

When radio became popular, Orson Welles employed it to tell his version of the H. G. Welles' novel "The War of the Worlds". Welles reported "on air" about an atrocious attack of extra-terrestrials starting on a farm in Grover's Mill, New Jersey and climaxing with a live report, describing giant Martian war machines releasing clouds of poisonous smoke across New York City. Although preceded by a clear introduction that the show was a fiction, it became famous for allegedly causing mass panic. But the reports about the extension of the panic and many actions of frightened people were as fake as the radio transmission.[10]

Since 1933, the Nazi regime used radio and film as tools to indoctrinate the population, to spread hate speech against Jews and other minorities in the country and to attack foreign governments. Competing governments did the same. In Britain, for example, the journalist Sefton Delmer created a fictional German figure called the "Chef" who reported faked news about Germany.[11]

When the Convention concerning the 'Use of Broadcasting in the Cause of Peace' was signed in 1936, the 19 Member States, all Members of the League of Nations, agreed to stop and prohibit all types of transnational broadcasting which "... is likely to harm good international understanding by incorrect statements" and that, in times of crises, broadcasters should only transmit information concerning international relations "the accuracy of which shall have been verified".[12]

## III. Fake News and the Internet

For some time now, we can observe that unregulated speech creates a real danger for democracy and specifically for free elections and referenda. Of course, the old world of gatekeepers has also never been free of one-sided or even fake news. But at least the public generally knows or can find out who is responsible for the reported facts and ideas.[13] Newspapers have reputations for being right or left

9 Great astronomical discoveries lately made by Sir John Herschel, L.L.D. F.R.S. &c. at the Cape of Good Hope, called the great moon hoax.

10 The Telegraph, May, 6th, 2016, https://www.telegraph.co.uk/radio/what-to-listen-to/the-war-of-the-worlds-panic-was-a-myth.

11 https://www.smithsonianmag.com/history/fighting-nazis-fake-news-180962481.

12 LNTSer 80; 186 LNTS 301.

13 Under German law each newspaper is obliged to publish an imprint, which contains information about the publisher, all journalists of the newspaper and the person responsible for a specific article.

wing, or for being serious or sensationalist. Traditionally there is good and bad journalism, and people are able to categorize specific commentaries and have developed standards to evaluate the truth of specific information.[14]

In contrast, where news is published by non-traditional media on the Internet, the reliability of the information and their sources are often difficult to evaluate. News on the Internet can be published by private individuals or groups who lack the means or even the goodwill to check their information. In other cases, the information published originates from foreign states or agencies which have a range of agendas and intentions, with fake news becoming an issue of increasing concern.[15]

## IV. The use of bots as free speech?

But, can bots like journalists in newspapers or broadcasting stations rely on free speech? Is an automat free or is it dependent on its master's will? Is then the person or the company who writes the bots program the one who can rely on fundamental rights of communications?

As the algorithms progress, there is an evolution towards generators of entire texts rather than just the completion of sentences. For example, the text generator GTP-3 is an artificial intelligence developed by *Open AI*, a business financed by Elon Musk and Microsoft, which can create text content when fed with single words or sentences.[16]

The program already created a fairy tale, starting with just one sentence, and created twitter posts on various topics. This illustrates the dangers mentioned above regarding fake news: sometimes, it is not possible tell the difference between machine-produced texts and another written piece by a human journalist. Can they be then also protected by the fundamental right to free speech?

Although these questions remain unsolved to a great degree until this day, the arguments in favor of the applicability of fundamental rights for journalism bots are not convincing.

Artificial intelligences are algorithms, which need to be programmed by human beings. In order to learn and create their own content, these algorithms

---

14 Fran Yeoman, 'The value of professional journalism', *The Independent* (4 December 2013) <https://www.independent.co.uk/voices/comment/the-value-of-professional-journalism-8982792.html>.

15 Nick Anstead and Ben O'Loughlin, 'Social Media Analysis and Public Opinion: The 2010 UK General Election' (2015) 20 *Journal of Computer-Mediated Communication* 204.

16 A Research paper on GTP-3 can be found via https://arxiv.org/pdf/2005.14165.pdf.

are trained with millions of texts serving as examples. Their learning process can be launched only by analyzing already existing data serving as a basis. This means, rather than being a genuine process of creation, the text generator can be compared with a statistic program. In this way, it does not express an opinion of its own.

This is illustrated by experiments with the GTP-3 bot, showing that the bot created a surprising percentage of anti-Semitic or sexist content, just because such content was made available to it beforehand.[17] On the other hand, it cannot produce content on new questions, for example to topics like COVID-19. This shows that, even though artificial intelligences are nowadays very advanced technologies, they are still dependent on existing data and do not create entirely new content, meaning they cannot be regarded as authors or journalists themselves.

Another aspect to look into is the general ability of computer program to be subjects of fundamental rights, specifically regarding their ability to form and express an opinion. Opinions are generally defined by being based on decision-making and underlying motivations.[18] In order to form an opinion, it is therefore necessary to be able to justify own actions and to act on moral standards. This is only the case for human beings, not machines (unless you take into account Robots such as Wall-e or other movie stars, who are still fictional).

Comparing the question to the application of fundamental rights to legal persons does not produce a different result. In that case, not the legal person itself is protected, but the collective of human beings behind it. Machines and algorithms cannot be held responsible, only the human beings who participated in the production. In addition, fundamental human rights are generally derived of human dignity, which cannot be applied to machines.

In consequence, fundamental rights, such as the freedom of speech and the freedom of press, are not applicable to content created by complex journalism bots such as GTP-3. It would therefore be possible for a legislative body to regulate the use of journalism bots without regard to the restrictions of fundamental rights.[19]

So, if it is not the robot which is at the source of the opinion, what is?

Two possibilities can be considered: the attribution of the opinion to the programmer or a journalist as the user of the algorithm.

17 https://twitter.com/an_open_mind/status/1284487376312709120.
18 Palasí/Spielkamp,AlgorithmWatch paper No. 4, Berlin 2017, p. 6.
19 Habel, Roboterjournalismus, 2019.

The algorithm itself as a product of his work can be attributed to the programmer. But the attribution of an opinion is more than a simple causality of action to output: it requires a subjective relation of the speaker to his statement. This can oftentimes be the case, even with the use of complex algorithms where the result is seemingly random, in case the programmer has control over the output by influencing the source code in a way that only certain results are possible. He can do this by including tendencies (for example: ranking criteria, the inclusion or exclusion of certain content).[20] Thereby, we can say that "code is law" in a sense that all product of bots is already regulated and biased.[21]

On the other hand, when used in journalism, texts will undergo a review before being published. If the journalism bot is only used for some phases of the journalistic process – when the fact basis is provided by the media business and the program only generates a text out of the data inserted – the journalists only uses the bot as a tool to refine, publish and spread the information. In this case, we can compare the usage of the journalism bot to other publishing tools. In this case, the journalist using the journalism bot would be the person disseminating an opinion. Furthermore, artificial intelligences tend to personalize to the user, e.g. the journalist producing texts on a regular basis. By accepting or dismissing text products, the bot will change his way of functioning in order to comply with the demands of the individual user.

Another possibility is to consider that the author of the text to be the media platform publishes it in their own name. Here, we can apply the general principle in media law that an opinion of a third person is regarded as coming from the person who published it via their own platform, unless they clearly dissociate themselves from the content.

## VI. Google "autocomplete" and the German Bundesgerichtshof

Another example for a complex algorithm are search engines such as google. Not only do they rank information from third persons according to the underlying source code, they also create their own content.

In 2013, the German Federal Court of Justice (Bundesgerichtshof) had to deal with the question whether Google was responsible for violations of personal

20 ibid.

21 Lawrence Lessig, Code Is Law: On Liberty in Cyberspace, https://www.harvardmagazine.com/2000/01/code-is-law-html.

rights created by the autocomplete-function of their search engine.[22] The most spectacular case was Bettina Wulff, the wife of Germany's former Head of State, Christian Wulff.[23] When entering only the first three letters of Mrs. Wulff's first name, the autocomplete function already created links to articles pretending that she was working for an "escort" service in Munich in her former life, meaning she was a prostitute.

Google's autocomplete function is based on an algorithm which includes, among others, the number of other user's search requests. Starting as a gossip to damage her husband's political reputation, a huge number of articles, social media discussions and other "information" sources spread in the Internet, although there was and is absolutely no evidence at all.

Of course, in the first instance, the dissemination of these fake news creates liabilities. However, most of the sources were anonymous.

The German Federal Court of Justice has ruled that it is primarily the conduct of the user that can lead to a violation of personal rights, such as the general right of personality, through untrue factual allegations. The predictions, which interfere with the general right of personality, are dependent on a certain user conduct.

As the Court correctly stated, the defendant cannot be blamed for using the software that creates the predictions, because this is an occupation which is protected by fundamental rights.

However, in the Court's opinion, the defendant is not acting in a mere technical, automatic or passive manner, because he processes the search items of the users with his own software in order to create predictions of the autocomplete function. Therefore, the Court concluded that the defendant was responsible for this process. The Court specified that the defendant could basically be reproached because he had not taken the necessary precautions to avoid that the predictions of the auto complete function would infringe the rights of third parties.[24] Aspects of possibility and reasonableness are especially important for the question of responsibility in cases of omission.[25]

22 BGH, Decision of 14 May 2013, Case no.VI ZR 269/12 = Multimedia und Recht (MMR) 2013, pp. 535–540 added with notes = epd medien 31 (2013), pp. 14–19.

23 The court did not decide the case, however the legal problems of the decision remain relevant.

24 BGH, Decision of 14 May 2013, Case no. VI ZR 269/12, § 26 = MMR 2013, pp. 535–540, 537 = epd medien 31 (2013), pp. 14–19, 18.

25 BGH, Decision of 14 May 2013, Case no. VI ZR 269/12, § 27 = MMR 2013, pp. 535–540, 537 = epd medien 31 (2013), pp. 14–19, 18.

The Court thus focuses on the question liability rather than looking at it from a constitutional perspective. Now, what can be concluded for the fundamental right of freedom of speech?

## VII. Regulations on the International Level

As we have seen, filters and ranking algorithms create problematic effects including the creation of 'filter bubbles'[26] by search engines and social networks as well as the so-called search engine biases. These phenomena seemingly interfere with the user's rights of access to information, which are being steered in a certain direction without anyone noticing. Multiple political players on national, European and international level are therefore initiating regulation to address these problems in order to determine the liability of content provided by Internet platforms, for example. Regarding search engines, solutions discussed include the creation of a public service search engine or to enforce obligation transparency regarding their code and their *modus operandi* to guarantee their neutrality.

It must be taken into account that search engines play a dominant role in today's world as they provide access to a very large flood of information, which had previously been unknown. Thus, they play a positive role of communication in the digital world and have a large impact on opinion building of the users. As an actor in the communication field, we cannot ignore them when discussing the fundamental right of freedom of speech. In a media context, we can draw a comparison with the figure of the "public watchdog" function of the press.

Their position is somewhat ambivalent: on the one hand, search engines have an institutional function by providing access to information and can thus be addressed by the users as subjects of fundamental rights; on the other hand, they are themselves affected by state regulation.[27]

So, can they invoke the right to freedom of speech against this legislation?

When regarding the applicability of fundamental rights of communication to search engines like google, not only the freedom of speech, but also the freedom of the press and the freedom of broadcasting have to be considered. The freedom of speech might thus be subordinate to the freedom of broadcasting.

The automated ranking provided by Google is not the basis for a "battle of arguments" generally required for the freedom of speech. But taking into

26 See for example Bakshy/Messing/Adamic, Science 2015, 1130, 1130.
27 Hartl, Suchmaschinen, Algorithmen und Meinungsmacht, 2019, p. 242.

account the principle of convergence, the freedom of broadcasting seems to be most fitting.

Whilst the European Court of Justice considered the fundamental rights of communication not to be applicable in the judgement *Google Spain,*[28] it can be considered to apply Article 11 of the Charter of Fundamental Rights of the European Union and Article 10 of the European Charter of Human Rights.

## VIII. Conclusion

Robotic speech is part of the opinion-building process and is not a mere product of coincidence, but is influenced by the source code of the bot as well as the behavior of the user.

When applying fundamental rights of communication to the programmers and providers of robots, regulation must comply with constitutional rules, specifically regarding the proportionality of the stipulations. Since also the users are protected by fundamental rights, and many providers of online services have monopoly status, regulation for robots should guarantee their transparency and neutrality in terms of the produced content.

28 European Court of Justice, Judgment of 13th of May 2014 – C-131/12 – Google Spain, § 85.

Ines Gillich, Rostock

# "Algorithmic Journalism" and Freedom of the Press

## I. Introduction

Freedom of expression is a fundamental right and a core value in the democratic process. It ensures that everyone is able to discuss, exchange, and debate ideas with others without censorship or reprisals.[1] Since this right applies only to human communication, it is questionable whether its ambit also extends to communication generated by an algorithm based on programmed parameters.

This article analyses whether the so-called "algorithmic"[2], "robot"[3] or "automated"[4] journalism is protected by the freedom of the press and whether and to which extend it may be subjected to state regulation.

---

1 For example, Article 5 para 1of the German Basic Law ("Grundgesetz") lists in several ways of expressing opinions, including freedom of the press and broadcasting, and prohibits censorship. Article 11 of the Declaration of the Rights of Man and of the Citizen of 1789, which is annexed to French Constitution states: "The free communication of ideas and of opinions is one of the most precious rights of man. Any citizen may therefore speak, write and publish freely, except what is tantamount to the abuse of this liberty in the cases determined by law." The First Amendment to the US Constitution provides that: "Congress shall make no law respecting an establishment of religion, or prohibiting the free exercise thereof; or abridging the freedom of speech, or of the press; or the right of the people peaceably to assemble, and to petition the Government for a redress of grievances." Article 10 para 1 of European Convention on Human Rights (ECHR) does also not distinguish explicitly between different communication rights, but refers only to freedom of expression.

2 C.W. Anderson, Towards a sociology of computational and algorithmic journalism, New Media & Society 15 (7), pp. 1005–1021.

3 Christer Clerwall, Enter the robot journalist: Users' perceptions of automated content, Journalism Practice 8 (5), pp. 519–53.

4 Matt Carlson, The Robotic Reporter, Digital Journalism 3 (3), pp. 416–431.

## II. Distinguishing "Algorithmic Journalism" from Other Forms of Modern Journalism

"Algorithmic journalism" uses AI-driven tools to generate coherent and fluent news content. It must be distinguished from other forms of modern journalism.

### 1. Aiding Software for Journalistic Activities

First of all, if a journalist merely employs assisting software for journalistic activities this is not considered "algorithmic journalism". For decades, journalists have been supported by computer programs that propose modifications, shortenings or changes to existing texts as well as regarding its layout. These tools help to adapt the texts to the specific formal and stylistic requirements of various news channels. The decision whether or not to accept the suggestions made by the algorithm, however, remains entirely with the human journalist.

### 2. User Behavior Analytics Software

Secondly, "algorithmic journalism" must also be distinguished from user behavior analytics programs (UBA). User behavior analytics (UBA), such as provided by Google Analytics, tracks, collects and assesses quantitative and qualitative user data and activities through monitoring, with the aim of providing information about which articles or passages and headings are of particular interest to visitors at a certain webpage.

### 3. Internet Bots

Finally, the use of "internet bots" is not "algorithmic journalism" in this context. An internet bot is a software application that runs automated tasks over the Internet. These tasks are typically simple and performed at a much higher speed compared to human Internet activity. Chat bots are algorithms designed to converse with a human user, and social bots are algorithms designed to mimic human behaviors to converse with patterns similar to those of a human user.

## III. Defining "Algorithmic Journalism"

Algorithms no longer merely support journalists in their work; instead, they carry out autonomous tasks by researching, editing, publishing and/or distributing information according to specified variables to specific recipients without human intervention in the process after the final programming or editing.

## 1. The Emerge of "Algorithmic Journalism"

Automatic narration is an emerging field of application. AI algorithms are employed to automatically convert facts derived from data silos and AI tools into readable stories without human involvement and within a fraction of a second. Such algorithms were first developed by a research team at Northwestern University in Illinois in 2010 with the aim of "teaching" algorithms to "understand" and "interpret" data and aiding journalists in producing text.[5] The result of this project, an algorithm called "Stats-Monkey", reported on baseball games so convincingly that the project was eventually incorporated into large business companies. One of these is Automated Insights ("Ai"). The company provides natural language generation (NLG) technology, based on an artificial intelligence system that uses silos of data, quantitative analysis and rules about style and good writing to produce news stories.[6] The algorithm can be programmed to write stories in any desired journalistic format, such as bullets, summaries, or long-form articles. These stories can be published in a variety of formats, such as emails, mobile applications, and the social media. For instance, in collaboration with the Los Angeles Times, Ai has developed an algorithm called "Quakebot" that generates texts for earthquake warnings. However, in 2017 "Quakebot" falsely reported an earthquake which, in fact, had already occurred in 1925. A few minutes later, the newspaper deleted the message – with the note that it was incorrect information. But the incident raised allegations that the robot was spreading "fake news".[7] The incident shows how important it is to maintain a correct data base from which automated news-generating tools derive their factual basis. In particular, an error-free process is crucial when it comes to generating articles about election or stock market results.

For instance, Finland's national public broadcasting company Yle's automated journalist "Voitto" not only generates articles about sports matches, but also about election results, and it creates graphics to be used by human journalists.[8] The long-established company "Associated Press" (AP) has been producing

5 Mercedes Bunz, The Silent Revolution – How Digitalization Transforms Knowledge. Work, Journalism and Politics Without Making Too Much Noise, 2014, p. 4.

6 See company details on www.automatedinsights.com.

7 Adrian Lobe, KI im Journalismus – Zu schnell für die Kontrolle, 26 July 2019, https://www.sueddeutsche.de/medien/ki-journalismus-fehler-1.4539688 (last visited 11 August 2021).

8 Leo Leppänen et al., in: Alonso et al. (eds.), Proceedings of the 10th International NLG-Conference 2017, 180 (192).

automated financial reports since 2014.[9] The Chinese Internet company Tencent started robotic journalism narrative writing activities in 2015, when it introduced its Dreamwriter algorithm, which is capable of writing a 1,000-word article in just one minute. In the Brazil 2016 Olympic Games, Dreamwriter produced 450 Olympic news items, mostly about China's domination of the sports, appearing just minutes after events ended.

These examples evidence that the automated evaluation and presentation of large amounts of data is a market with enormous potential.[10] Robot content writers are also being embedded into electronic games, producing real-time stories based on how the users are performing during the games.

## 2. How Algorithms Generate News

"Algorithmic journalism" is based on two pillars: Software that automatically extracts figures from huge data silos, and algorithms that automatically convert them into readable stories. In computer programming terms, an algorithm is a set of well-defined instructions to solve a particular problem. During the past years, enormous advances have been made due to new intelligent machine learning techniques.[11] Current applications for automated journalism range from simple code that extracts figures from a database, which are then used to fill in the blanks in pre-written template stories, to more sophisticated approaches that create more complex narratives. The latter relies on big data analytics and natural language generation technology.[12]

The structure of a journalism bot can be described in a simplified manner using an "input-throughput-output model".[13] In the first step, the software collects available data such as – in the case of football match – box scores, historical records or player demographics. The journalism bot is not (yet) able to independently question informants.[14] However, theoretically, with the help of a crawler, it is able to compile a knowledge base from data available on the Internet. In practice, pre-structured data is used, deriving either from third-party

9 Lobe, supra note 7.

10 Matt Carlson, The Robotic Reporter – Automated journalism and the redefinition of labor, compositional forms, and journalistic authority, Digital Journalism 2014, p. 416 (421).

11 Wolfgang Ertel, Grundkurs Künstliche Intelligenz, 4th ed. 2016, p. 313.

12 Andreas Graefe, Guide to Automated Journalism, 2017, p. 6 (16).

13 Konstantin Nicholas Dörr, Mapping the field of algorithmic journalism, Digital Journalism, 4 (6), p. 700 (703).

14 Graefe, supra note 12, at 6.

providers or directly from journalistic editorial offices. The pre-structuring of the raw data is important to ensure the quality of the output. It stands to reason that, e.g., weather data, sports data, financial data or data from forensic medicine have so far formed the basis of these applications, because this data is already pre-structured by the database manufacturer. The so-called knowledge base also includes a language base with prefabricated and fixed text elements, the "templates".[15] Moreover, the knowledge base can store empirical values, e.g. which words, expressions and information have already been processed into an output, or which information is particularly relevant for a specific recipient.[16] In addition, a digital memory can be implemented. For example, the results of previous elections can be saved, so that the journalism bot – if and as far as specified by the programmer – can draw a comparison with previous election results.

In the "throughput phase", the data is selected and processed.[17] Algorithms employ statistical methods to identify important and interesting events in the data. Those may include unusual happenings, a player's extraordinary performance, or the decisive moment for the outcome of a game. In the document structuring sentence planning phase, the length of a text and the abstract arrangement of the text elements and style issues are determined.[18]

Next, the software classifies and prioritizes the extracted data by importance and subsequently arranges the newsworthy elements by following predefined rules to generate a narrative. Finally, the story can be uploaded to the publisher's content management system, which can publish it automatically.[19]

## 3. Potentials and Dangers of "algorithmic journalism"

Research studies have shown that readers can hardly distinguish between articles written by human journalists and those written by an algorithm.[20] Whereas texts written by human beings are more likely to be read, the computer-written

---

15 Ehud Reiter/ Robert Dale, Building applied natural language generation systems, Journal of Natural-Language Engineering, 1997, p. 57 (63).

16 Ibid., explaining that, e.g. a business letter corpus can be based on real letters sent out in the past, for example; while a weather report corpus can be based on real reports written in the past.

17 Dörr, supra note 13, at 703.

18 Ibid., at 704.

19 Ibid.

20 Anja Wölker, Thomas E. Powell, Algorithms in the newsroom? News readers' perceived credibility and selection of automated journalism, Journalism, February 2018, SAGE Publications, https://doi.org/10.1177/1464884918757072 (last visited 11 August 2021).

texts, however, are generally considered more credible, since they usually contain more facts and figures.[21] Depending on their programming, algorithms can produce more unprejudiced results than the average human being. This raises the question of whether algorithms are possibly even better news writers than human journalists.

On the one hand, "algorithmic journalism" holds great potential. The promise is that it frees up capacities in the editorial offices: journalists no longer have to write reports under time pressure, but have time for more profound and investigative research.

On the other hand, "algorithmic journalism" bears the risk of manipulation and abuse. For instance, if an algorithm spreads a false report on stock market results, the stock prices could collapse. Experts have warned that the error rate in fully automated processes is high. This makes it necessary to clearly disclose responsibility and allocate legal liability. Although the structure of algorithms follows logical patterns, they are the product of human programming and its preconditions. Thus, they are not free from bias, since they encode the values and assumptions of their creators.[22] Hence, algorithms are only as impartial as the people who program them. What is more, hidden prejudices and bias can creep into algorithms unnoticed not only through programming, but also due to an inadequately selected data base. Moreover, the use of algorithms creates new societal risks for democratic societies. With the revelations about how Social Media platforms might have been used to influence previous democratic elections, the public realized that algorithmic decision-making can affect the democratic chances of participation as well as political equality of opportunity.[23]

Keeping the risks that arise from rule-based and learning systems in check is a challenging task for the legal system. The call for legislative action to regulate the use of algorithms especially in journalism is raised.[24] For most advanced algorithms applications, the user cannot see how they operate; underlying algorithms and their decision-making criteria thus remain a "black box". With

---

21 The communication and social sciences discuss the effects this technology has on the jobs of journalists, the organization of publishers, the recipients and ultimately on our democratic society as a whole. Cf. Dörr, supra note 13, at 700.

22 Mario Martini, Regulating Algorithms – How to Demystify the Alchemy of Code?, in: Martin Ebers et al. (eds.): Algorithms and Law, 2020, p. 100 (105).

23 John Herrman/ Mike Isaac, Conservatives Accuse Facebook of Political Bias, The New York Times, 9 May 2016, https://www.nytimes.com/2016/05/10/technology/conservatives-accuse-facebook-of-political-bias.html (last visited 11 August 2021).

24 See Martini, supra note 23.

regard to the use of journalism bots, the question arises as to which regulatory measures are necessary and advisable. In particular, it is demanded that the use of this technology must be transparent and that messages written by algorithms must be identified as such.

State regulation, however, must abide with the constitutional requirements of the right to freedom of communication. The constitutional framework determines which regulatory measures can or must ultimately be taken. Therefore, it is necessary to analyze first whether and to what extent "expression" by fully autonomous AI-driven tools for journalistic purposes is protected by the right to freedom of the press.

## IV. The Fundamental Right of Freedom of Communication

Freedom of speech is one of the oldest and most precious human rights. It is an essential part of the human as a social being, since it is human nature to communicate with each other.

### 1. National and International Guaranties of a Free Press

Freedom of speech and other communication freedoms, such as freedom of press or broadcasting, are enshrined as fundamental rights in a number of national, European, international and regional instruments. In the US, the First Amendment stipulates that "Congress shall make no law . . . abridging the freedom of speech, or of the press." Use of the single word "expression" to reach speech, press, petition, association, and the like, raises the question of whether the free speech clause and the free press clause are coextensive, or whether one reaches where the other does not. In particular, it is questionable whether the "institutional press" is entitled to greater freedom from governmental regulations or restrictions than are non-press individuals, groups, or associations. Justice Stewart has argued "[t]hat the First Amendment speaks separately of freedom of speech and freedom of the press is no constitutional accident, but an acknowledgment of the critical role played by the press in American society. The Constitution requires sensitivity to that role, and to the special needs of the press in performing it effectively."[25] According to Chief Justice Burger "[t]he Court has not yet squarely resolved whether or not the Press Clause confers

25 Houchins v. KQED, 438 U.S. 1, 17 (1978) (concurring opinion). See also First National Bank of Boston v. Bellotti, 435 U.S. 765, 798 (1978) (Chief Justice Burger concurring).

any freedom from government restraint upon the 'institutional press', which is not enjoyed by all others."[26]

In Germany, Article 5 para. 1 of the Basic Law ("Grundgesetz") guarantees freedom of the press together with freedom of expression, freedom of broadcasting, and freedom of information. Together, they can be referred to as freedom of communication. But freedom of the press is not a general special case of freedom of expression. In addition to the editor and journalist, publishers, editors and wholesalers are also protected as well as the accountant and advertising manager. Freedom of the press protects in particular to press-specific activities, the final product, its prerequisites and framework conditions.[27]

Communication freedoms are also protected by Article 10 para 1 ECHR. In its interpretation this provision, the European Court of Human Rights (ECtHR) has held that "freedom of expression constitutes one of the essential foundations of [democratic] society, one of the basic conditions for its progress and for the development of every man".[28]

## 2. Democratic Function of a Free Press

Overall, a free press facilitates the free flow of information and of opinions. In this respect, the press not only conveys opinions and information between the elected representatives of the people and the recipients, but acts as a "public watchdog"[29]; it takes on a social task: Through speech and contradiction in the context of a public discussion, opinions are formed as a "preliminary stage" for the formation of political will in parliament.[30]

## 3. Positive Obligations for the State

All fundamental communication guaranties vouch for subjective defense rights against state interference.[31] Freedom of the press, in addition, includes an

---

26 First National Bank of Boston v. Bellotti, 435 U.S. 765, 798 (1978). Chief Justice Burger concluded that the institutional press had no special privilege as the press.

27 BVerfGE 85, 1 (12); 86, 122 (128).

28 ECtHR, Handyside v. the United Kingdom, Application no. 5493/72, Judgment 7 December 1976, at para. 49.

29 See e.g. ECtHR, Lingens v Austria no 9815/82, judgment of 8 July 1986, para 44; Jersild v Denmark no 15890/89, judgment of 23 September 1994 [Grand Chamber], paras 31 and 35.

30 See BVerfGE 20, 162 (174).

31 Ulrich Karpen, Freedom of expression as a basic right: A German view, The American Journal of Comparative Law, Spring, 1989, 37 (2) (Spring, 1989), p. 395 (396).

objective legal aspect: the state's duty to protect for the press. For the ECHR, the ECtHR has held that these positive obligations imply, among other things, that the States are required to establish an effective mechanism for the protection of authors and journalists in order to create a favorable environment for participation in public debate of all those concerned, enabling them to express their opinions and ideas without fear.[32] The ECtHR determined that Member States have a positive obligation "to create a favorable environment for participation in public debate by all persons concerned."[33] The creation of a favorable environment means, e.g., that the State is obliged to protect a journalist against attacks by people who feel insulted by his publications. It does not necessarily include an environment in which news users or media organizations are protected against deleterious effects of AI-driven tools on freedom of expression unless AI-driven tools would threaten the individual rights of journalists, news media, and other participants to public debate. The European Council has further elaborated in various recommendations and publications on the nature of such a favorable environment. It includes, among others, the need to provide the media with both financial and non-financial support and to protect them from threats.[34]

## 4. The Subjective Relation between the Communicator and the Recipient of an Utterance

All communication freedoms relate to human communication, since human action is based on social interaction. A subjective attitude of the person making the statement to the subject of the statement must be recognizable. At best, algorithms may simulate this interaction, but they cannot generate it. Unlike human beings, machines and algorithms cannot be assigned moral and legal responsibility for their actions. Since thoughts originate in the human mind, the question arises whether this natural association is lost through the use of a journalism bot that "forms" the "opinion" or "statement" of the programmer and/or the journalist who uses the algorithm.

32 ECtHR, Dink v. Turkey, Appl. No. 7124/09, Judgment, 14 September 2010, para 137.

33 Ibid.

34 Declaration on the financial sustainability of quality journalism in the digital age, 2019, DECL (13/02/2019)2 (declaration on the financial sustainability of quality journalism); draft recommendation on algorithmic systems, 2018, para. 7.1; recommendation on pluralism, 2018, para. 2.11; Recommendation of the Committee of Ministers to Member States on the protection of journalism and safety of journalists and other media actors, 2016, CM/Rec(2016)4 (recommendation on the safety of journalists,), para. 18, 38.

### (a) Attribution

The question is whether the utterance made by a journalism bot can be attributed to a person. In this regard, the predictability and traceability of the automation are crucial factors.

Journalism bots are programmed with decision trees and text modules so that the selection decisions and the form of presentation are abstractly defined by the programmer.

The selection of the topic, for example, the coverage of football games, is based on the assessment that recipients are interested in these games. Moreover, the vocabulary and the stylistic form can also be programmed as desired. In this respect, journalism bots are software with tendencies and bias.

### (b) Protection of a Special Political Leaning or Ideology

The guarantee of the freedom of the press includes protection for media companies which pursue a specific ideological tendency. The bearers of this right can freely determine the type, orientation, content, and form of a publication. A journalistic or technical bias may therefore appear problematic in terms of communication studies, but under legal aspects, the right of tendency is closely rooted in freedom of expression and the media.

If journalism bots are used in the private media sector, the owner of the respective media company and not an instructed, employed or commissioned programmer has the authority to determine the tendency of the articles produced by journalism bots.

Even though the programmer cannot foresee the ultimate result in detail ex ante, he or she exerts influence over the factors programmed in the algorithm. This makes the algorithmic result attributable to the programmer.

### (c) "Internal" and "external algorithmic journalism"

With respect to the personal scope of human rights protection by the freedom of the press, a distinction may be drawn between "internal" and "external algorithmic journalism".[35] When media companies develop journalism bots in-house with the aim of publishing reporting content – so-called internal algorithmic journalism,[36] they are directly bearers of the right to freedom of the press. In particular, freedom of the press or media should not be restricted to groups of

35 Dominic Habel, Roboterjournalismus, 2019, p. 98.
36 Ibid., p. 102.

people who are already active in the media sector. It also includes the establishment of media companies with innovative business models.

In the case that algorithms are programmed by third companies ("external algorithmic journalism"),[37] those software companies can also invoke the right to freedom of the press if they develop and make journalism bots available to a publishing media company. This is possible because freedom of the press also extends to all natural and legal persons who participate in the preparation, production, or distribution of reporting articles.[38] Software companies can therefore invoke freedom of the press because the programming activity is directly related to the content of the articles published by the publishing media companies.

Whether the level of protection for software companies decreases is determined by the type and intensity of the participation in the journalistic process. The extent to which the software company retains control over the editing of the articles depends on whether the media company can program the journalism bot on its own. If the journalism bot is only made available as a download and the software company retains the general maintenance and optimization of the software inhouse, the nature and intensity of the relationship to the media company's reporting activities decrease. Just as Microsoft, a manufacturer of operating systems for computers, does not benefit from the protection of the freedom of press, software companies in this constellation cannot invoke this fundamental right.

### *(d) The Use of Bots as a Form of Deception about Identity*

The utterance of an opinion or a fact, however, is not limited to its content alone. The accompanying circumstances under which the utterance was made are also relevant. Therefore, the modalities of an utterance are of particular importance to establish a specific link between the communicator and the recipient.[39] Even though the persuasiveness of the argument is an essential factor, from the perspective of the recipient, it is also relevant who makes the utterance, and by which means and methods. Nevertheless, expressions made anonymously or with the use of a pseudonym are legally protected under the right to free speech.[40]

---

37 Habel, supra note 35, at 98.

38 See Christian Starck/Andreas Paulus, in: v. Mangoldt/Klein/Starck (eds.), GG, Vol. 1, 7th ed. 2018, Article 5, para. 133.

39 See e.g. BVerfGE 93, 266 (295).

40 Federal Court of Justice (BGH), ruling of June 23, 2009, VI ZR 196/08, para. 38 (BGHZ 181, 328) with regard to anonymous expressions.

With regard to bots, however, it may be appropriate to differentiate between two scenarios: If a bot user pseudonymizes or anonymizes the generated utterances, the recipient is basically aware of how to deal with this. However, if the bot pretends to be a human being, the bot user is essentially deceiving about the special connection between the content of the opinion and the person behind it. In the latter case this expression may be subject to stricter legal restriction and regulation, as will be explained below in further detail.

## V. Legal Standards for a Regulatory Framework

National legislators find an arsenal of conceivable measures at their disposal, which can be applied preventively parallel to the use of the software applications, and ex post in the form of damages and legal protection. In addition, state regulation can be supported by accompanying measures of professional self-regulation.

As explained above, regulative measures by the State find their legal basis in positive obligations to protect and to guarantee communication rights vis-à-vis threats stemming from private individuals or groups. Regulative measures have to take into account the bot users' freedom of communication on the one hand and the public interest in the integrity of a free and democratic decision-making process as well as the subjective rights and interests of the individuals in their roles as recipients and competitors in an unadulterated communication process on the other. Thereby, the regulator must take into account that the use of algorithms can be detrimental to the communicative equality of opportunity if direct human communication were to be repressed. The use of algorithms poses challenges for the maintenance of a diverse media landscape. Unequal access of small versus large media organizations to such tools and data could lead to a less favorable environment for some companies, which can affect their economic viability. Large news companies often have access to better tools and more data silos, which enables them to more easily provide users with relevant news.[41] Therefore, the state can be obliged to create a communication regime that safeguards these conflicting interests.[42]

---

41 See e.g. Recommendation of the Committee of Ministers to Member States on the roles and responsibilities of internet intermediaries, 2018, CM/Rec(2018)2 (recommendation on intermediaries), para. 7.

42 The Council of Europe has issued recommendations that explore the responsibilities of Member States and private parties with regard to AI-driven tools and big data. See e.g. Guidelines to Convention 108 on Artificial Intelligence and data protection, 2019, T-PD(2019)01; Guidelines to Convention 108 on the protection of individuals with regard to the processing of personal data in a world of Big Data, 2017, T-PD(2017)01;

## 1. Duties and Responsibilities of Journalists

The increased protection afforded to the press as a "public watchdog" is subject to the condition that journalists comply with the duties and responsibilities coming along with this public function. Article 10 para. 2 ECHR makes this clear when stipulating that the exercise of this freedom carries with it "duties and responsibilities".

The scope of a journalist's duties and responsibilities depends on the situation and the technologies used for communication.[43] These duties and responsibilities apply to journalists as well as for other actors that contribute to public debate, including owners or publishers of news outlets[44] and online news portals.[45]

In the light of the foregoing, it must be concluded that the use of algorithms to deliver news could also come with specific duties and responsibilities. To determine the applicable duties and responsibilities of such news media, the potential impact of the medium is an important factor that must be taken into account.[46]

Such duties and responsibilities include "acting in good faith and on an accurate factual basis and [to] provide "reliable and precise" information in accordance with the ethics of journalism."[47] Media professionals producing or publishing content which violates the ethics of journalism might thus not be protected by the freedom of press. The Council of Europe and the ECtHR prescribe a differentiated and graduated approach to the legal principles applicable to different media actors. For instance, the ECtHR has held that the legal responsibilities of print and audiovisual media may differ from the responsibilities of online media.[48]

---

Recommendation of the Committee of Ministers to Member States on Big Data for culture, literacy and democracy, 2017, CM/Rec(2017)8; Declaration by the Committee of Ministers on the manipulative capabilities of algorithmic processes, 2019, Decl (13/02/2019)1; Draft Recommendation of the Committee of Ministers to Member States on human rights impacts of algorithmic systems, 2018, MSI-AUT(2018)06.

43 Handyside v. the United Kingdom, Appl. No. 5493/72, Judgment of 7 December 1976, para. 49.

44 Sürek v. Turkey (No. 1), Appl. No. 26682/95, Judgment of 8th July 1999, para. 63; Editions Plon, Appl. No. 58148/00, Judgment of 18 May 2004, para. 50.

45 Magyar Tartalomszolgáltatók Egyesülete and Index.hu Zrt v. Hungary, Appl. no. 22947/13, Judgment of 2 February 2016, para. 62.

46 Jersild, Appl. No. 15890/89, Judgment of 23rd September 1994, para. 31.

47 Fressoz and Roire, Appl. No. 29183/95, Judgment of 21st January 1999 [Grand Chamber], para. 54.

48 Delfi AS v. Estonia, Appl. No. 64569/09, Judgment of 16th June 2015 [Grand Chamber], para. 113.

## 2. Algorithmic Journalism as a Journalistic Activity

The legal term "journalism" is to be understood in a functional sense.[49] It does not only include the work of journalists in the formal sense, i.e. by people who have been trained in journalism, work for a newspaper or broadcasting company and/or are members of a journalists' association. Instead, it applies also to a person who regularly publishes articles on matters of public interest to an indefinite group of people and observes certain minimum standards of responsible research and publication.[50] Therefore, not only traditional radio programs and newspaper articles, but also blogs and algorithm-based publications can be "journalistic".

Journalistic activity, however, requires editorial responsibility.[51] Editorial responsibility means exercising effective control over both the composition of content and its delivery. The criterion of "editorial responsibility" distinguishes content providers from mere information intermediaries.

### *(a) Duty to Correctness and to Truthfulness*

The duty of truthfulness for journalists relates to facts.[52] In contrast to opinions that are shaped by the subjective relationship of the person making the statement into an object of expression, and therefore cannot be right or wrong, facts are characterized by an objective relationship between the object of expression and reality. Logically, an obligation to report truthfully can only relate to the factual basis of an expression of opinion. The obligation to truthfulness requires checking the news for content and origin with due diligence prior to distribution.

With respect to the use of algorithms for journalistic purposes, specific care must be taken when gathering the factual basis and when programming the parameters of the algorithm, as well as with regard to verification and transmission.[53]

49 Jan Oster, European and International Media Law. 2017, p. 10.

50 Ibid., p. 11.

51 Jan Oster, Media Freedom as a Fundamental Right, 2015, p. 57.

52 See e.g. BVerfGE 12, 113 (130) emphasizing the media's duty to report truthfully.

53 Mark Hansen et al., Artificial Intelligence: Practice and Implications for Journalism. New York: Tow Center for Digital Journalism, Columbia University, https://doi.org/10.7916/D8X92PRD, p. 8 (last visited 11 August 2021).

### (b) Duties and Responsibilities which Relate to Editorial Decision-Making

The automation of news and the introduction of AI-driven tools to newsrooms also raise questions regarding the scope of the duties and responsibilities of the editors. Tasks that were previously performed by human journalists and editors are delegated to algorithm-based software applications. It also raises the question of who is responsible to ensure that robot journalism adheres to journalistic duties and responsibilities.

Arguably, media actors have a responsibility to develop professional rules regarding the risks of AI-driven tools for bias and media diversity. There is a need to stipulate clear conditions of responsibility and editorial oversight regarding automated processes. A clear conceptualization of editorial control in the context of AI-driven tools is necessary to identify when automated journalistic products qualify as journalism at all, and which actors should oversee and bear editorial responsibility.

However, existing journalistic codes and mission statements regarding editorial responsibility are still focused on traditional journalistic routines.[54] Algorithm-based tools are more complex. They are elements of potentially far-reaching structural changes in internal routines and provide challenges with respect to the division of responsibility between human beings and machines. The automation of journalistic functions and the disruptive structural changes that accompany these processes raise new legal and ethical challenges, which must be addressed by professional journalistic codes. In particular, journalistic algorithmic ethics would need to address questions such as how to deploy algorithmic tools in journalism responsibly according to notions of fairness, balance and diversity. This makes it necessary to create internal checks and balances to detect, for example, bias or lack of diversity in recommender systems, make processes transparent and thus controllable, and clearly assign editorial responsibility for automated recommendations and news products.

## 3. Legal Regulation of "algorithmic journalism"

State regulation of algorithmic journalism should not try "to crack the regulatory walnut with a sledge-hammer".[55] Not every software application and not every

54 Recommendation of the Committee of Ministers to Member States on self-regulation concerning cyber content, 2001, Rec (2001)8 (self-regulation and user protection against illegal or harmful content on new communications and information services); Recommendation of the Parliamentary Assembly on Ethics of journalism, 1993, 1215 (1993).

55 Martini, supra note 22, at 111.

machine-learning algorithm poses a threat to fundamental rights justifying regulation. All regulatory efforts must thus pursue a legitime regulatory aim and they must abide to the principle of proportionality.

### *(a) Ensuring Communicative Equality and the Freedom to Receive Information*

Ensuring communicative equality is a legitimate regulatory goal.[56] For the German Federal Constitutional Court, it is conceivable that the State has a protective duty to ward off dangers that could arise for a free press system from the formation of monopolies of opinion.[57] Algorithmic journalism may change the quality and type of news that users receive. It is argued that social network sites and search engines, powered by AI-driven tools, have in particular formed a news system that disadvantages traditional journalism.[58] The broad application of algorithms for news production, pushing back traditional human-written news, could, therefore, lead to news users having less access to the type of (quality) journalism that forms the core of the "public watchdog" function of the media, such as stories that uncover misbehavior of politicians and corporations, or stories that give voice to minorities.[59] As a result of their positive obligation, States may have to ensure that news organizations do not limit news users' right to be informed or to hold an opinion by using algorithms. Moreover, such positive obligations may also extend to the quality of the news that news users receive via algorithm-based systems.

### *(b) Protection against Fully Automated News Production*

Since the use of algorithms to produce news stories has come along with personal data collection, it interferes with the rights to privacy and personal data

56 Wolfgang Schulz, Gewährleistung kommunikativer Chancengleichheit, 1998, p. 168; ECtHR, Manole and Others v. Moldova, Appl. No. 13936/02, Judgment of 17th September 2009, para. 101 (holding that Member States are obliged to ensure that citizens are able to receive balanced and diverse news).

57 BVerfGE 20, 162 (176).

58 Emily J. Bell et al., *The platform press: How Silicon Valley reengineered journalism*. New York: Tow Center for Digital Journalism, Columbia University, https://academiccommons.columbia.edu/doi/10.7916/D8R216ZZ (last visited 12 August 2021).

59 Council of Europe, Committee of Ministers, Declaration on the financial sustainability of quality journalism, 2019, paras. 7 and 9 on the pressure on journalists to produce sensationalist content.

protection. The possibility of access to data on individual reading patterns collected by newsrooms and internet intermediaries to create or improve AI-driven tools poses a latent threat to individual rights and the democratic process.

On the level of the European Union, one step toward the regulation of algorithms has been taken by the General Data Protection Regulation (GDPR).[60] Article 22 para 1 of the GDPR gives data subjects the right of defense within particularly sensitive automated decision-making procedures. However, the normative radius of this provision is limited. It establishes numerous exceptions and refers to "a decision based solely on automated processing.", therefore covering merely decisions being made without any substantial human influence. Thus, Article 22 GDPR does not completely solve regulatory tasks for the majority of algorithmic decision-making procedures.

The Council of Europe has demanded that individuals should be informed about all algorithmic decision-making that concerns them, and they should have meaningful control over these processes, including and have access to effective remedies for violations of human rights by private actors that design, develop or deploy algorithmic systems.[61]

### *(c) Transparency Obligations for Algorithm-Based News Aggregators*

As long as communication can be traced back to a human being, algorithmic journalism is, as explained above, a technical modality of an utterance. But there is a qualitative difference between communication mediated by software and direct interpersonal communication.

Transparency obligations have long been recognized in the field of classic journalistic mass communication, such as in the form of imprint obligations. Such imprint obligations serve to create legal liability. They ensure that third

60 Regulation (EU) 2016/679 of the European Parliament and the Council of 27 April 2016 on the protection of natural persons with regard to the processing of personal data and on the free movement of such data, and repealing Directive 95/46/EC (General Data Protection Regulation).

61 Council of Europe, Committee of experts on human rights dimensions of automated data processing and different forms of artificial intelligence – Addressing the impacts of Algorithms on Human Rights, Draft Recommendation of the Committee of Ministers to Member States on the human rights impacts of algorithmic systems, para. 5.1. This draft recommendation on the human rights impacts of algorithmic systems proposes a horizontal set of guidelines, directed at States, public and private sector actors, with a view to promoting an environment of legal certainty in which both human rights and innovation can thrive.

parties affected by the reporting can defend their rights against the reporting media. Furthermore, concentration of power in the media area becomes visible through general disclosure obligations.

In the context of algorithmic journalism, the question arises not only whether the existing transparency obligations are applicable, but also whether – de lege ferenda – more far-reaching transparency obligations are permissible. It is argued that news aggregators profoundly influence opinion-forming processes. Given the importance for the democratic aspect of a free and pluralistic public discourse, these services should be subject to stronger transparency requirements than other algorithm-based services if they reach a widespread audience.[62]

Various points of reference for transparency obligations are conceivable. Transparency can relate to the disclosure of the automation per se, to the decision variables, such as the abstract program structure and trend values, to the source code or to the raw data source. Transparency obligations collide with the legitimate secrecy interests of media companies, which are protected by the freedom of the press.[63] They must be balanced against each other according to the principle of proportionality.

A law that obliges media companies to disclose the raw data, however, would be disproportionate. Editorial secrecy is a necessary condition of free, independent press. In particular, government agencies are not allowed to gain insight into the research work and the editing of articles by viewing documents, contacts and editorial processes. It is obvious that this can undermine the press' public watchdog function. The programmer of an algorithm which is used to write an article must not necessarily be made transparent as long as the media company behind the article can be clearly identified.

A less intensive interference with the guarantee of free press would be an obligation to formally separate the human communication mediated by journalism algorithms from direct human communication by labelling obligations. In this regard, a regulatory concept stemming from data protection law could be transferred to robot journalism. Article 15 para. 1 lit. h GDPR provides that the data subject shall have the right to obtain from the controller confirmation as to whether or not personal data concerning him or her are being processed, and, where that is the case, access to the personal data and information on the existence of automated decision-making, including profiling and, at least in those cases,

---

62 Martini, supra note 22, p. 121.

63 Tarleton Gillespie, The Relevance of Algorithms, in: Gillespie et al. (eds.), Media Technologies: Essays on Communication, Materiality, and Society, 2014, p. 167, 185.

meaningful information about the logic involved, as well as the significance and the envisaged consequences of such processing for the data subject. Transferring this regulatory concept to algorithmic journalism, it could be required that those employing journalism bots have to render the abstract program structure and the trend values implemented in the program transparent. This does not lead to a full disclosure of the source or program code, which is protected by copyright law, but enables recipients to understand the decision-making process of the program. A labeling requirement does not interfere with the right to publish an article anonymously. As long as the fact that a contribution was generated fully automatically by means of a journalism bot is made transparent, the programmer can remain anonymous. Moreover, the programmers would only have to pre-formulate a template with the relevant information, which can be automatically published with the article itself.

## VI. Conclusion

"Algorithmic journalism" as defined here is a reporting activity, thus enjoying protection under the right to freedom of the press. The programming of the source code of the algorithm is a form of opinion formation, because it is where individual assessments are made and trends and bias of the programmer are processed. The person who writes the algorithm determines the language of the program, the vocabulary, the sentence structure and the tonality. Even though it is not possible to foresee how the program will decide and how exactly the text will literally be composed, the final result is attributable to the human programmer.

National and international human rights instruments protect the forming, expressing, and disseminating of facts and opinions, which is important for democratic processes. Especially a free press serves a democratic function as a "watchdog of democracy". However, fundamental protection by the freedom of press also comes along with – in the words of Article 10 para. 2 ECHR – to special "duties and responsibilities". The decisive factor here is whether or not algorithm-based journalism can meet the journalistic due diligence requirements applicable for anthropogenic journalism. National legislators are under a positive obligation to maintain and guarantee a free communication order. To safeguard this, the State may regulate the use of algorithms for journalistic purposes.

Russell L. Weaver, Louisville

# Robotic Speech and the First Amendment

The history of free expression is closely linked to advances in communications technology.[1] Johannes Gutenberg's development of movable type, which led to development of the printing press in the fifteenth century, was the first major technological advance.[2] Previously, people had limited communications possibilities.[3] Manuscripts existed, but they were written in long hand, usually by monks who functioned as scribes rather than as creators of original content, and the manuscripts were written in Latin and about religion[4] (although universities began to create their own handwritten manuscripts by the thirteenth century).[5] The printing press revolutionized communication by making it possible for people to write in their own languages about diverse topics,[6] and led to a flowering of information which precipitated profound changes in government,[7] religion,[8] and science.[9]

---

1 *See* RUSSELL L. WEAVER, FROM GUTENBERG TO THE INTERNET: FREE SPEECH, ADVANCING TECHNOLOGY AND THE IMPLICATIONS FOR DEMOCRACY (Carolina Academic Press, 2nd ed., 2019).

2 *See* IRVING FANG, A HISTORY OF MASS COMMUNICATION: SIX INFORMATION REVOLUTIONS 26 (1997).

3 *See* FROM GUTENBERG TO THE INTERNET, *supra* note 1, at 5–7.

4 *See* A HISTORY OF MASS COMMUNICATION, *supra* note 2, at 24 (For "nearly a thousand years, from the fall of the Roman Empire to the introduction of the printing press, the monasteries kept the sputtering flame of knowledge. Monks, hunched over their scriptorium desktops, painstakingly hand lettered and painted manuscripts magnificently to produce books, *incunabula*, for libraries and cathederals.").

5 *See* FROM GUTENBERG TO THE INTERNET, *supra* note 1, at 6–7.

6 *See id.*, at 11.

7 *Id.*, at 14–18. Indeed, the U.S. Declaration of Independence, which implicitly rejects the Divine Right of Kings, and explicitly endorses democracy (the document specifically declares that "the power to govern derives from the consent of the governed"), ultimately owes a debt to Gutenberg. Although the document was written by Thomas Jefferson, the ideas reflect arguments articulated by European philosphers. Undoubtedly, Jefferson learned about those ideas through printed books. Id., at 114–16.

8 *Id.*, at 13–14.

9 *Id.*, at 13.

The Gutenberg press remained the dominant method of communication for many centuries until electricity was harnessed in the seventeenth century.[10] In the ensuing centuries, electricity led to the development of a number of new communications technologies, including the telegraph, radio, television, cable, satellite communications, and the internet.[11] Electricity was revolutionary because it allowed information to move much more quickly than people could move.[12] Whereas it had previously taken more than a week to send a message across the U.S.,[13] electricity made it possible to transmit a message in a matter of seconds.[14] Radio expanded the communication possibilities even further by making it possible to send sounds over long distances.[15] During World War II, President Franklin Delano Roosevelt used the radio to communicate with the U.S. population through his so-called "Fireside Chats,"[16] and Edward R. Murrow broadcast live reports regarding the bombing of London.[17] Television made it possible to transmit, not only sounds, but also video.[18] The full power of television was on display during the assassination of President John F. Kennedy.[19] And

---

10 *Id.*, at 39.

11 *Id.*, at 39–46 & 61–65.

12 *Id.*

13 *Id.*

14 *Id.*

15 *See* David Crowley & Paul Heyer, Communications in History: Technology, Culture, Society 204 (5th ed., 2007).

16 *See* Matt Bai, *Pushing the Presidential Message Into the Broadband Age, The New York Times* A12 (Jan. 25, 2011).

17 *See* A History of Mass Communication, *supra* note 2, at 164 ("Night after night from the rooftop above his studio, Murrow's microphone picked up the sounds of anti-aircraft fire as Murrow painted word pictures of the bombing. Across the Atlantic in a nation still at peace, listeners eagerly awaited his CBS Radio reports, which helped to shift American sentiment from the isolationist wish to avoid war to a feeling of wanting to come to the aid of the beleaguered English populace.").

18 *See id.*, at 165 ("The murder of the President (John Kennedy) under mysterious circumstances, followed by the murder of his alleged assailant, both shown as they happened, led not to angry citizens taking to the streets, but to a citizenry glued to television sets to watch the drama that unfolded over more than three days. All programming, all commercials, and all other news vanished. Television news pictures hypnotized viewers as night and day unfolding events were told and shown over and over.").

19 *Id.* ("Attitudes changed on November 22, 1963, the day of the assassination of John F. Kennedy. The murder of the President under mysterious circumstances, followed by the murder of his alleged assailant, both shown as they happened, led not to angry

satellite technology made it possible to beam information around the world.[20] CNN reporters, holed up in a Baghdad hotel room, used satellites to provide live coverage of the outbreak of the first Persian Gulf War to CNN for broadcast on television.[21]

The internet has been as revolutionary as the Gutenberg printing press.[22] The internet was transformative because it enabled ordinary people to mass communicate.[23] Before the internet, most advanced communications technologies came with "gatekeepers," the owners or operators of communications technologies, who were able to control the use of those technologies.[24] Ordinary people could not access printing presses (or, for that matter, radio or television) without the permission of those gatekeepers.[25] Gatekeepers also controlled cable and satellite systems.[26] The internet has largely been free of gatekeepers. Although social media companies exercise control over their own platforms,[27] the internet is a remarkably resilient technology that allows most people to freely communicate without interference from gatekeepers.[28] Thus, even though many social media companies have banned certain individuals and groups (e.g., the Daily Stormer, Alex Jones and Infowars),[29] all of those groups and individuals are readily accessible on the

---

citizens taking to the streets, but to a citizenry glued to television sets to watch the drama that unfolded over more than three days. All programming, all commercials, and all other news vanished. Television news pictures hypnotized viewers as night and day the unfolding events were told and shown over and over.").

20 *See id.*, at 169.

21 *See* Walter Goodman, *WAR IN THE GULF: TV CRITIC'S NOTEBOOK, On Television, the Theater of War, The New York Times*, at A19 (Jan. 17, 1991) ("Television achieved the excitement of old-time radio reporting last night as war began in the Persian Gulf.").

22 *See* From Gutenberg to the Internet, *supra* note 1, at 61–65.

23 *See id.*, at 62 ("The internet complimented the personal computer by providing ordinary individuals with an effective means for distributing the documents that they had created on their PCs.").

24 *See id.*, at 21–38 & 47–60. "Even though the Gutenberg printing press had a profound impact on the sweep of history, it's impact was limited by two considerations. First, many governments sought to repress the ability of private individuals to use the printing press. Second, the ability of ordinary individuals was limited by so-called "gatekeepers" who controlled access to the technology." *Id.*, at 21.

25 *Id.*, at 21–38 & 47–60.

26 *Id.*

27 *Id.*, at 124–132.

28 *Id.*, at 132–133.

29 *Id.*, at 129–130.

internet.[30]

Of course, the greatest strength of the internet – the fact that it enables ordinary people to freely communicate with each other – is also its greatest weakness.[31] Individuals can use the internet for a variety of nefarious purposes, including the distribution of "fake news"[32] and child pornography,[33] as well as to perpetrate fraudulent schemes,[34] and to disseminate hate speech.[35] For example, individuals have used the internet to promote claims that vaccines cause autism, and that the moon landing, as well as the shootings in Parkland, Florida, and Las Vegas, Nevada, were hoaxes.[36]

This article focuses on one particular problem with the internet: the participation of bots, and other forms of robotic speech, to participate in online discussions. As we shall see, bots are flooding the internet with comments, Facebook "likes," and other forms of communication.[37] The question is whether robotic speech should be treated as "free expression" that is protected under the First Amendment to the U.S. Constitution.

## I. The Rise of the Bots

"Bots" have become increasingly commonplace on the internet. Indeed, one commentator claims that there has been a "'Cambrian explosion' of robotics technologies"[38] with millions and millions of bot accounts and bot messages on the internet.[39] One estimate suggests that there are as many as 48 million Twitter

---

30 *Id.*, at 132–133.

31 *Id.*, at 139.

32 *Id.*, at 139–155.

33 *See* Spencer S. Hsu, *Website selling 'real' rape and child pornography videos shut down after arrest in Netherlands, Justice Department says, The Washington Post* (Mar. 12, 2020).

34 *See* Elisabeth Leamy, *Don't Fall For These Latest Cyber Scams, The Washington Post* (Mar. 30, 2017). https://www.washingtonpost.com/lifestyle/home/don't-fall-for-these-latest-cyber-scams/2017/03/29/810f9dfe-0f12-11e7-9d5a-a83e627dc120_story.html

35 *See* From Gutenberg to the Internet, *supra* note 1, at 139–170.

36 *See* Bryan Casey & Mark A. Lemly, *You Might be a Robot*, 105 Cornell L. Rev. 287, 322 (2020).

37 *Id.*

38 *See* Casey & Lemly, *supra* note 36, at 295.

39 *See* Ashley Fox, *Automated Political Speech: Regulating Social Media Bots in the Political Sphere*, 18 First Amendment L. Rev. 114 (2020); Michael Newberg, *As Many as 48 Million Twitter Accounts Aren't People, Says Study*, CNBC (Mar. 10, 2017, 1:09 PM),

accounts that are run by bots rather than human beings, and which account for as much as 25 % of all tweets in the 2016 presidential election.[40] A second estimate concludes that there as many as 91 million fake accounts on Facebook, and 24 million bots on Instagram.[41] A Pew Research Center study suggested that nearly two-thirds of all tweeted links were shared by bots had links to news and current events.[42] At one point, "YouTube had as much traffic from bots masquerading as people as it did from real human visitors."[43]

The explosion of bot communications was dramatically illustrated by a rulemaking proceeding conducted by the U.S. Federal Communications Commission (FCC) regarding whether to repeal its net neutrality requirement. During that rulemaking, the FCC received an absolutely extraordinary number (23 million) of public comments.[44] When the comments were analyzed by a machine learning engineer, the analysis suggested that at least 1.3 million comments were submitted using stolen or misused identities, and that more than 445,000 comments came from Russian or German email accounts.[45] Many fake comments were submitted by bots which were detectable because they involved duplicate comments differentiated only by the substitution of a synonym for a particular word.[46]

As technology has advanced it has become increasingly difficult to distinguish human beings from bots. And, in a curious twist, "we have not only seen bots pretending to be human . . . but an increasing number of humans pretending to be bots."[47] If we try to assess the level of public support for a proposition or an issue, bots can deceive us by suggesting that particular ideas have far more (or far less) support than they actually have.

---

https://www.cnbc.com/2017/03/10/nearly-48-million-twitter-accounts-could-be-bots-says-study.html; *see also* Jack Nicas, *Does Facebook Really Know How Many Fake Accounts It Has?*, N.Y. Times (Jan. 30, 2019), https://www.nytimes.com/2019/01/30/technology/facebook-fake-accounts.html

40 *Id.*, at 117.

41 *Id.*

42 *See* Stefan Wojcik, *5 Things to Know about Bots on Twitter*, Pew Research Center, Factank (Apr. 9, 2018).

43 Michael H. Keller, *The Business of Serving Up YouTube Views: Streams Are for Sale, Eyes Not Included, The New York Times* A18 (Aug. 12, 2018).

44 *See* Jason Tashea, *No Comment: The FCC*, ABA Journal 33 (Aug. 2018).

45 *Id.*

46 *See* Brian X. Chen, *The Internet Trolls Have Won. Get Used to It, The New York Times* B-7 (Aug. 9, 2018).

47 Casey & Lemley, *supra* note 36, at 291.

## II. Bots and the First Amendment

It is tempting to assume that bot speech should be denied First Amendment protections. The U.S. governmental system is grounded in democratic principles. As Thomas Jefferson emphatically stated in the Declaration of Independence, "the power to govern derives from the consent of the governed."[48] If the U.S. is committed to democratic principles, then freedom of expression is the engine that makes democratic governance work. As the U.S. Supreme Court has recognized, "speech concerning public affairs is more than self-expression; it is the essence of self-government."[49] Or, as the Court stated in *Citizens United v. Federal Election Commission*,[50]

> "Speech is an essential mechanism of democracy, for it is the means to hold officials accountable to the people. The right of citizens to inquire, to hear, to speak, and to use information to reach consensus is a pre-condition to enlightened self-government and a necessary means to protect it. The First Amendment has its fullest and most urgent application to speech uttered during a campaign for political office. It is inherent in the nature of the political process that voters must be free to obtain information from diverse sources in order to determine how to cast their votes."[51]

The Court has also declared that "maintenance of the opportunity for free political discussion to the end that government may be responsive to the will of the people and that changes may be obtained by lawful means, an opportunity essential to the security of the Republic, is a fundamental principle of our constitutional system."[52]

Of course, at the time that the First Amendment was written and ratified, computers and robotic speech did not exist. Although the Gutenberg printing press did exist, most modern communications technologies (e.g., radio, television, cable, satellite and the internet) did not. As a result, in formulating and ratifying the First Amendment, eighteenth century Americans would have been focused on the ability of actual people to participate in and influence the political

---

48 U.S. Declaration of Independence (July 4, 1776).

49 *Garrison v. Louisiana*, 379 U.S. 64, 74 75 (1964); *see also R.A.V. v. City of St. Paul*, 505 U.S. 377, 423 (1992) (Blackmun, J., concurring) ("core political speech occupies the highest, most protected position.").

50 558 U.S. 310, 339 (2010).

51 *See also Virginia v. Black*, 538 U.S. 343, 365 (2003) ("Political speech, of course, is 'at the core of what the First Amendment is designed to protect.' ").

52 *New York Times Co. v. Sullivan*, 376 U.S. 254, 269 (1964) *(quoting Stromberg v. California*, 283 U.S. 359).

process through oral speech or through printed works created by hand or (more likely) using the Gutenberg press. So, the question is how the First Amendment should be applied to modern technologies like bots which were beyond the contemplation of the founding generation.

Some have argued that bot speech, even if it involves "bot-created content," should be protected under the First Amendment because of its communication value. As one commentator noted, "If the courts focus on the publisher, it would be difficult for AI communicators to receive journalistic protections. If the courts focus on what was published, however, AI communicators have a better chance of succeeding, particularly if their content can be seen as a public good."[53] Another commentator makes a similar argument: "Even though bots are abstract entities, we might think of them as having free speech rights to the extent that they are promoting or promulgating useful information for the rest of us."[54]

The problem with this approach is that it attempts to draw a clear dividing line between human speech and bot speech, a line that does not exist. In some instances, what seems like bot speech may actually be the speech of human beings.[55] For example, in one case, a Ted Cruz tweet included an embedded button that allowed the receiver to instantly and easily retweet the original Cruz message.[56] A machine controlled when the new tweet would actually be sent.[57] Thus, even though the retweets might have appeared to be bots, because Cruz's original message was retweeted in mass without alteration, the tweets involved the conscious decisions of actual persons who were choosing to retweet Cruz's message.[58]

The Cruz incident illustrates the fact that it can sometimes be perilous to attempt to distinguish between human speech and machine speech. The reality is that much speech occurs through machines of one sort or another. Since the

53 Jared Schroeder, *Are bots entitled to free speech?*, Columbia Journalism Review (May 24, 2018). https://www.cjr.org/innovations/are-bots-entitled-to-free-speech.php

54 Matt Field, Do social media bots have a right to free speech?, Bulletin of the Atomic Scientists (Jan. 9, 2019) (quoting Professor Laurent Sacharoff). https://thebulletin.org/2019/01/do-social-media-bots-have-a-right-to-free-speech/#:~:text=%E2%80%9CEven%20though%20bots%20are%20abstract%20entities%2C%20we%20might,free%20speech%20right%2C%20almost%20independent%20of%20its%20creators.%E2%80%9D

55 *See* Casey & Lemly, *supra* note 36, at 289–290.

56 *Id.*, at 289–290.

57 *Id.*, at 290.

58 *Id.*

1400s when Gutenberg invented the printing press, machines have played a vital role in human communication. Radio, television, satellites and the internet all involve machines that help make mass communication possible. Radio enabled FDR to use Fireside Chats to communicate directly with the public and to do so on a grand scale. In a similar vein, the internet enables ordinary people to mass communicate and widely disseminate their ideas. When bots flood social media with ideas, their messages will sometimes reflect the views of a single individual who is simply trying to influence the public debate. In other words, in some instances, bots and the internet provide nothing more than a mechanism by which individuals can distribute their ideas.

A distinction could perhaps be made between bots that simply relay the speech of actual persons, and bots that create their own content. One commentator suggests that many bots function as "content communicators" (in the sense that they share existing content) rather than functioning as "content creators."[59] However, there are some bots, controlled by artificial intelligence, that actually do generate their own content. These bots can "automatically create tweets without direct human oversight,"[60] and can "spread information, and misinformation, with alarming speed."[61] They can "like," "share," "follow," or "retweet" in conjunction with an online account.[62] Moreover, bots that use functions such as "like" or "share" are less likely to reveal their non-human nature "by inadvertently using poor grammar or issuing off-color remarks."[63] In this way, bots can create "floods of emotive reactions to headline news" that appear real.[64]

But, even in the instance of "content-creator" bots, the dividing line between bots and people is not clear-cut. Bots never function completely autonomously. On the contrary, someone programs the bots to intervene in public debates on specific issues, and programs the bots to give a particular type of response. So, even "content-creator" bots involve a mix of human conduct and bot conduct.

59 *See* Fox, *supra* note 39, at 117–118 (quoting a study by Jared Schroeder).

60 *See* Wojcik, *supra* note 42.

61 Ryan Calo & Madeline Lamo, We Need To Decide If Robots Are Protected By The First AmendmentShould bots be required to tell you that they're not human? And would such a rule be a violation of free speech?, BuzzFeedNews (July 26, 2018).

62 Casey & Lemly, supra note 36, at 322.

63 *Id.*, at 322.

64 *Id.*

## III. The Problem of "Outsized" Influence

Some might object to bots on the ground that they allow individuals who use them to have an outsized influence on the public debate by flooding social media and other online venues with their comments and views. As one commentator noted, "Thanks to the democratization of computer science know-how, state-sponsored 'influence campaigns,' and a profound cheapening of computing power, this category of spoofer is more prolific than ever. Even actors of limited means can now use bots to generate thousands of public comments, put celebrity faces on pornographic videos, inflate friend and follower lists, or maliciously influence the electoral process."[65] As a result, bots can help someone "weaponize" speech by making it appear that certain ideas have much greater support and acceptance than they actually have.[66] "Bots have been used to mislead users; to artificially inflate follower counts, likes, and retweets; and to manufacture consensus on divisive issues. In 2015 and 2016, bots were frequently used to make topics trend on Twitter, creating the false impression that certain stories were of great importance."[67]

The difficulty is that the problem of outsized influence has always existed. Historically, the U.S. media has been controlled by rich individuals or corporations. At one time in the U.S., prominent families controlled the newspaper business, including the Ochs-Sulzbergers (New York), Binghams (Louisville), Chandlers (Los Angeles), Medills (Chicago and elsewhere),[68] and Hearsts (various cities).[69] Those who controlled the means of communication were able to bias the news in the direction of their preferred positions. For example, during World War II, the Medills advocated for an isolationist position that some attributed to their dislike for President Franklin Delano Roosevelt.[70] William Randolph Hearst aggressively used his newspapers to campaign for causes that he supported,[71] as well as to

65 *Id.*, at 307.

66 Jonathan Peters, *The Modern Fight for Media Freedom*, 18 First Amendment L. Rev. 60, 68 (2020).

67 Renee Diresta, *A New Law Makes Bots Identify Themselves—That's the Problem*, Wired (July 24, 2019). https://www.wired.com/story/law-makes-bots-identify-themselves/

68 *See* Joseph Epstein, *Dynasts of the Daily Press*, reviewing Megan McKinney, *The Magnificent Medills: America's Royal Family of Journalism During a Century of Turbulent Splendor* (2011); Amanda Smith, *Newspaper Titan: The Infamous Life and Monumental Times of Cissy Patterson* (2011), *The New York Times* Book Review Section, at 17 (Oct. 16, 2011).

69 *See* W.A. Swanberg. Citizen Hearst 26, 40, 43–44, 47 & 100 (1961).

70 *See id.*, at 14.

71 *See id.*, at 100 & 119–127.

promote his preferred views and positions.[72] Prior to the American Revolution, there is evidence that printers similarly used their presses to pursue the interests and positions that they preferred. For example, there are stories about James Franklin (Benjamin Franklin's brother) and his opposition to positions taken by prominent Bostonian Cotton Mather who (to James) represented the Boston "establishment."[73]

If ordinary individuals can now use bots to leverage their own ideas, why should they be precluded from doing so? Should we privilege only the speech of rich people and corporations who control the traditional media? Even today, there is considerable evidence of media bias, and that includes media outlets who try to portray themselves as neutral and objective (e.g., National Public Radio).[74] So, why shouldn't ordinary individuals be allowed to use technology to help them level the playing field and to push their preferred positions? In some respects, bot technology might be regarded as simply a mechanism by which less wealthy individuals can widely disseminate their ideas. Since we are not prepared to limit the power of media outlets, it seems inappropriate to deprive ordinary people of their mass communication machines.

## IV. Disinformation

Some might object that bot speech can lead to the widespread dissemination of disinformation. In recent years, as previously noted, there has been a flood of false claims.[75] Bots have played a major role in the circulation of such information.[76] Indeed, it is relatively easy to find disinformation online. As one commentator noted, "digging up large-scale misinformation on Facebook was as easy as finding baby photos or birthday greetings."[77] In 2018, there "were doctored photos of Latin American migrants headed towards the United States border, as well as easily disprovable lies about the woman who accused Justice Brett

---

72 *See* Jerome A. Barron, *Access to the Media – 1967 to 2007 and Beyond: A Symposium Honoring Jerome A. Barron's Path-Breaking Article "Access Reconsidered,"* 76 Geo. W. L. Rev. 826, 832 (2008).

73 *See* H.W. Brands, The First American: The Life and Times of Benjamin Franklin 25 (2000).

74 *Id.*

75 *See* Diresta, *supra* note 67.

76 *Id.*

77 *See* Kevin Roose, *Facebook Had a Good Election, But It Can't Let Up on Vigilance, The New York Times*, The Shift B1 (Nov. 8, 2018).

M. Kavanaugh of sexual assault, cooked up by partisans with bad faith agendas."[78] Indeed, "every time major political events dominated the news cycle, Facebook was overrun by hoaxers and conspiracy theorists, who used the platform to sow discord, spin falsehoods and stir up tribal anger."[79]

Bots played a significant role in the 2016 presidential election as armies of bots spread false information on fake social media accounts.[80] Common Sense Media, a nonprofit focused on media and technology in relation to children, claims that Facebook and Twitter had a combined total of 100 million bot accounts, and that these accounts wreaked havoc online: "The destructive role of unlabeled Russian bots in the American election is well documented...Bots played a significant role in attacking high school students victimized by the Parkland, Florida, shooting, in encouraging kids to use electronic cigarettes, in spreading racism online, and in recent human rights violations in Kenya and Mexico." Online bots, which people misperceive as human, "lend false credibility" to various messages, California State Sen. Robert M. Hertzberg (D-Van Nuys) said, and are a major contributor to fake news.[81]

But the mere presence of disinformation does not provide an adequate basis for suppressing bot speech. The traditional media can sometimes be very biased in favor of its preferred positions,[82] but few would argue that government can or should regulate the media in order to eliminate that bias. On the contrary, in the U.S., we tend to assume that the remedy for bias or disinformation is responsive speech. As a result, there are few remedies for disinformation, and the U.S. Supreme Court has held that even false speech may be entitled to First Amendment protection.[83] Of course, the government may prohibit certain limited classes of false speech (e.g., perjury in judicial proceedings).[84] In addition, if an individual disseminates false and defamatory information about another person, it may be possible to recover for defamation.[85] However, it is extremely

---

78 *Id.*

79 *Id.*

80 Calo & Lamo, *supra* note 61.

81 Kristin Houser, Should Bots Have a Right to Free Speech? This Non-Profit Thinks So. Does the First Amendment apply to AIs, too?, Futurism (May 25, 2018).

82 *See* From Gutenberg to the Internet, *supra* note 1, at 53–58.

83 *See United States v. Alvarez*, 567 U.S. 709 (2012).

84 *Id.*

85 *See* Russell L. Weaver, Andrew Kenyon, David F. Partlett & Clive Walker, The Right to Speak Ill: Defamation, Reputation & Free Speech (Carolina Press 2006).

difficult for public officials to recover for defamation,[86] as well as for public figures to do so,[87] and defamation litigation is relatively uncommon in the United States.[88] In addition, courts rarely enjoin false speech, except when it involves false commercial speech, because injunctions against speech are regarded as "prior restraints" and therefore are presumptively unconstitutional.[89]

Under the First Amendment, government is generally prohibited from censoring speech simply because it regards that speech as untruthful (or, for that matter, "offensive" or "distasteful").[90] Perhaps the American attitude is a reaction to early European attempts to repress speech, as well as to similar attempts in the British Colonies.[91] In any event, the U.S. does not have "truth commissions" or "censorship boards" which are allowed to dictate which ideas are permissible, and which are not. On the contrary, the U.S. Supreme Court is wary of governmental attempts to control the flow of information, and has generally regarded both content-based and viewpoint-based restrictions on speech as presumptively unconstitutional.[92] It is not for government to dictate what people should believe, but rather for the people to decide for themselves.

Some commentators have attempted to analyze fake news under the marketplace of ideas metaphor, and have suggested that restrictions on speech might be appropriate in an effort to avoid distortions to the marketplace.[93] For example, one scholar stated that to "suggest that the proliferation of fake news represents a breakdown in a market where consumers trade ideas is to presume that fake news concerns ideas about which the public can debate. It doesn't. Fake news does not concern ideas. It concerns facts."[94] Of course, there are lots of problems with this analysis. First, it is not clear that there really is a "marketplace of ideas" in the Holmesian sense that free and open debate can be expected to lead society to "truth."[95] Moreover, even as to facts, the U.S. does not prohibit or sanction the

---

86 *See New York Times Co. v. Sullivan*, 376 U.S. 254 (1964).

87 *See Curtis Publishing Co. v. Butts*, 388 U.S. 130 (1967).

88 *See* The Right to Speak Ill, *supra* note 85.

89 *See New York Times Co. v. United States*, 403 U.S. 713 (1971); *Near v. Minnesota*, 283 U.S. 697 (1931).

90 *See Cohen v. California*, 403 U.S. 15 (1971); *Tinker v. Des Moines Independent Community School District*, 393 U.S. 503 (1969).

91 *See* From Gutenberg to the Internet, *supra* note 1, at 21–32.

92 *See R.A.V. v. City of St. Paul*, 505 U.S. 377 (1992).

93 *See* Ezra Waldman, *The Marketplace of Fake News*, 20 U. Pa. J. Const. L. 845 (2018).

94 *Id.* at 847.

95 *See* C. Edwin Baker, *Scope of the First Amendment Freedom of Speech*, 25 U.C.L.A. L. Rev. 964 (1978).

articulation of false facts in most contexts. Unlike some countries (*e.g.*, France) which make it criminal to deny the truth of certain facts (*e.g.*, the French Gayssot law criminalizes denial of the Holocaust),[96] the First Amendment does not generally allow the U.S. government to formally declare that certain facts must be accepted (or, at least, cannot be denied) on pain of criminal penalty.[97] Indeed, if government were allowed to declare "truth," there is a risk that the government might be biased in the sense of declaring "facts" to be true because they favor one side of the political debate and disadvantage the other side (e.g., the Obama administration might have declared that climate change is a fact while the Trump administration might have declared the opposite). So, the reality is that the marketplace of ideas does not play a meaningful role in U.S. First Amendment analysis except as a reflection of the idea that all ideas and information must be allowed into the "marketplace" without censorship Under such circumstances, it is difficult to preclude "fake news." How can one reliably determine what news is "fake?" More to the point, is the U.S. society prepared to give government that authority? Presumably not.

## V. Possible Legislative Responses

One avenue for controlling bot speech is through legislation. For example, Congress enacted the B.O.T. Act in 2016 which makes it a crime to use bots to make bulk purchases of tickets for entertainment events.[98] However, the purchase of entertainment tickets is at the periphery of the issues raised in this paper.

In 2018, California passed the Bolstering Online Transparency Law (BOT Law) which requires that a bot disclose its identity (as a bot) in any online situation (including social media) involving a commercial transaction or an attempt to influence a vote in an election.[99] The sponsor of the law explained its intent as follows: "The heart of this issue is simple: we need to know if we are having

96 *See* Russell L. Weaver, Nicholas Delpierre & Laurence Boissier, *Governmentally Imposed Truth: An Examination of France's Holocaust Denial Law*, 41 Texas Tech. U. L. Rev. 495 (2009).

97 *See id.*

98 Cal. Bus. & Prof. Code §§ 17940–17943.

99 § 17941 provides as follows: "(a) It shall be unlawful for any person to use a bot to communicate or interact with another person in California online, with the intent to mislead the other person about its artificial identity for the purpose of knowingly deceiving the person about the content of the communication in order to incentivize a purchase or sale of goods or services in a commercial transaction or to influence a vote in an election."

debates with real people or if we're being misled."[100] According to the sponsor, the law was proposed as a way to " 'protect democracy' from disingenuous accounts programmed to misinform the public."[101] He added that: "Bots continue to misrepresent public sentiments and perceptions about topics, or to mute dissenting opinions and distract from current events."[102]

When it enacted the law, the California legislature heard evidence that, during the 2016 political campaign, "social media bots accounted for 3.8 million (19 percent) of US election-related tweets. Twitter told a Senate committee that a Russian intelligence agency had 2,752 Twitter accounts, of which 47 percent were bots."[103] "To its supporters, the bot law is a respite in a time of online chaos, a sorely needed defense against the disinformation, misinformation, and filter bubbles that exist on today's sprawling social media platforms. For regulators, it's a chance to do something—to make social media manipulation less of a free-for-all—as the US heads into another election."[104]

It is not clear that the BOT Law can or will be effective in regulating bot speech. One problem is that the law suffers from vagueness, and may be unconstitutional. For one thing, it is difficult to know what constitutes an attempt "to influence a vote in an election." In *Minnesota Voters Alliance v. Mansky*,[105] the U.S. Supreme Court struck down a Minnesota law that made it a crime to "display campaign material, post signs, ask, solicit, or in any manner try to induce or persuade a voter within a polling place or within 100 feet of the building in which a polling place is situated" to "vote for or refrain from voting for a candidate or ballot question." While the Court agreed with Minnesota that "some forms of advocacy should be excluded from the polling place,"[106] it struck the law down because it was too vague. Minnesota construed the law as banning "only words and symbols that an objectively reasonable observer would perceive as conveying a message about the electoral choices at issue in the polling place."[107] But, the Court found it difficult to determine "What qualifies as an "issue."[108] The

---

100 *Should Bots Have a Right to Free Speech?*, *supra* note 81.

101 *See* Mark Rapazzini, *California's Online BOT Bill*, Heffler Claims Group (July 20, 2020). https://www.hefflerclaims.com/general/california-online-bot-bill/

102 *Id.*

103 *Id.*

104 Diresta, *supra* note 67.

105 136 S.Ct. 1876 (2018).

106 *Id.*, at 1887.

107 *Id.*, at 1889.

108 *Id.*

Court viewed the state's answer ("any subject on which a political candidate or party has taken a stance") as too broad: [109] As the Court noted, "Candidates for statewide and federal office and major political parties can be expected to take positions on a wide array of subjects of local and national import,"[110] and therefore the restriction was unduly vague.[111]

If the law had been limited to "promoting a group with recognizable political views" concerning the election at hand, the Court concluded that such a limitation would "make matters worse."[112] As the Court noted,

> Any number of associations, educational institutions, businesses, and religious organizations could have an opinion on an "issue confronting voters in a given election." For instance, the American Civil Liberties Union, the AARP, the World Wildlife Fund, and Ben & Jerry's all have stated positions on matters of public concern. If the views of those groups align or conflict with the position of a candidate or party on the ballot, does that mean that their insignia are banned? See *id.*, at 24, n. 15 (representing that "AFL–CIO or Chamber of Commerce apparel" would be banned if those organizations "had objectively recognizable views on an issue in the election at hand"). Take another example: In the run-up to the 2012 election, Presidential candidates of both major parties issued public statements regarding the then-existing policy of the Boy Scouts of America to exclude members on the basis of sexual orientation.[6] Should a Scout leader in 2012 stopping to vote on his way to a troop meeting have been asked to cover up his uniform?[113]

California's BOT Law suffers from as many, if not more, problems than the Minnesota law that was struck down in *Mansky*. What does it mean to "influence" an election? Is one attempting to do so if one criticizes a public official (or, for that matter, praises the official), or must one specifically advocate for the election or defeat of an official? "Is sharing legitimate news stories or voting locations a form of 'influence?' "[114] Is someone attempting to influence an election if he/she

---

109 *Id.*

110 *Id.*, at 1899–1900.

111 *Id.*, at 1900:

See, *e.g.*, Democratic Platform Committee, 2016 Democratic Party Platform (approved July 2016) (stating positions on over 90 issues); Republican Platform Committee, Republican Platform 2016 (approved July 2016) (similar). Would a "Support Our Troops" shirt be banned, if one of the candidates or parties had expressed a view on military funding or aid for veterans? What about a "# MeToo" shirt, referencing the movement to increase awareness of sexual harassment and assault? At oral argument, the State indicated that the ban would cover such an item if a candidate had "brought up" the topic. . . .

112 *Id.*

113 *Id.*

114 Diresta, *supra* note 67.

makes no reference to a particular candidate or candidates, but praises or is critical of issues on which a candidate (or candidates) have taken explicit positions? Undoubtedly, the California law suffers from all of the same problems evident in the Minnesota law.

But there is another major flaw in the California law: the law defines the term "bot" as "an automated online account where all or substantially all of the actions or posts of that account are not the result of a person."[115] What does that mean? When bots mass distribute information, does that necessarily make it bot speech? At the very least, the law is extremely vague. In the Ted Cruz example, cited above, bots were involved in retweeting the original Cruz Tweet. But, in each instance, the retweet occurred because a human being clicked on the embedded link in order to make that happen. Presumably, that situation meets the definition of something for which "all or substantially of the actions" involve that of a person, and therefore would be exempt from the law. Should it matter that the retweets were sent to a machine which decided when and whether to forward them? One would think not.

The law also would seem to be inapplicable when one or more people use robotic technology to widely disseminate their own ideas. If the bots are simply a platform for forwarding the ideas of individual people, then presumably the requirement that the post come from (or substantially all of it from) a human is satisfied, and it would again be offensive to label such speech as nothing more than "bot speech."

Presumably, the California law is directed towards bots that are acting relatively autonomously to distribute information broadly on platforms like social media. In other words, suppose that a bot is programmed to automatically respond when it spots the discussion of particular issues (e.g., vaccines, black lives matter, or other "hot button" matters), and to give a pre-programmed response. Would this type of action be regarded as bot speech under the law? Even here, the distinction between "bot speech" and "human speech" is not so clear. One commentator asks whether it is possible to "imagine any online activity--no matter how automated--that is 'not the result of a [real or corporate] person' at the end of the day."[116]

Suppose that an individual has strong feelings about vaccines, or for that matter, about the BLM movement, and programs computers to respond with his position each time someone posts on these topics. Given the human involvement,

---

115 Cal. Bus. & Prof. Code § 17940(a).

116 Casey & Lemley, *supra* note 36.

would the bot responses be regarded as bot speech or as human speech? Since bots rarely act with complete autonomy, and are usually programmed by individuals to respond on issues that the individuals care about, and to state the preferred positions of those individuals, it is not clear that they fall within the scope of the California law.

To make matters more complicated, at times, there can be interaction between the bots and their human operators. As one commentator noted,

> Malign bots are often automated for some percentage of the time, whether that's intraday or over their life cycle, particularly if the operator wants to "age" the account so that it doesn't look like it was created for one purpose. But the account is then taken over by a human operator when it's time to execute an influence operation. Russia's infamous Internet Research Agency accounts, for example, are often carelessly referred to as "bots." In reality, their most effective and well-known personas were largely human-operated. Persona accounts run directly by people are a much more daunting problem than purely automated bots when it comes to mitigating online influence operations.[117]

Indeed, Russia's Internet Research Agency (IRA) employed so-called "trolls" (essentially, Russian individuals) who posed as Americans and weighed in on controversial issues.[118] These "trolls" purportedly numbering in the thousands, worked 12 hour shifts, and were prepped regarding what to say on U.S. social media and on how to say it.[119] Each troll was expected to produce at least 80 comments per day and to have at least 20 shares per day.[120] Once a troll created a post, he would forward it to one of a "countless" number of fake accounts in an effort to create large numbers of "page views."[121] Although some Clinton supporters believe that the Russian efforts tipped the election,[122] it is not clear how much impact these posts had on the U.S. electorate.[123] There were lots of

---

117 Diresta, *supra* note 67.

118 *See* Mike Issac, *For Social Media, an Election Day Test, the New York Times, The Week in Tech* B3 (Nov. 12, 2018); Anton Troianovski, *A former Russian troll speaks: "It was like being in Orwell's World, The Washington Post* (Feb. 18, 2018).

119 *See* Neil MacFarquhar, *Unrest Spreads, Some Violently, in Middle East, The New York Times*, at A1 (Feb. 17, 2011).

120 *Id.*

121 *Id.*

122 *See* Jonathan Martin & Maggie Haberman, *Moscow's Hand Swirled in U.S., but Whether It Tipped Election is Unclear, the New York Times* A11 (Feb. 19, 2018) (As Clinton's campaign communications director alleged, "Russia succeeded in weakening her [Clinton] enough so that the Comey letter could knock her off.").

123 *See id.* ("And the nation's intelligence agencies say they do not have any way to calculate whether the Russian effort swung the election.").

problems with the Clinton campaign, including Clinton's general unpopularity.[124] As one commentator observed regarding the posts, "the audience seemed to grow more jaded and paid less attention to what they wrote."[125]

There are some potential limitations on bot speech. For example, the U.S. government prohibits foreign interference in U.S. elections, and it can prosecute foreigners who attempt to interfere. However, whether the U.S. government can get hold of interfering foreigners (e.g., Russian trolls) is far from clear. In addition, social media platforms can attempt to aggressively control bots. As private entities, they are not constrained by the First Amendment, and they have greater freedom to regulate bot speech. But, as noted, these companies can control only their own platforms, and bots can roam more freely on the rest of the internet. Moreover, there are so many bots that it is not clear that social media companies can effectively identify and control them.

## Conclusion

The internet has ushered in a brave new world of communication which is distinguished from earlier technologies by the fact that it has enabled ordinary people to mass communicate on a scale that was previously unheard of. At the same time, by giving everyone the ability to mass communicate, the internet has created opportunities for individuals to spread disinformation, as well as child pornography, hate speech and other objectionable content.

Bots have only added to the problem. As people are inundated with information, they face difficulties trying to evaluate that information. As the FCC rulemaking on net neutrality illustrates, bots are capable of flooding the internet with automated speech. Not only can they inundate rulemakings, they can overwhelm social media accounts. In addition, they can send messages, do "likes" of postings, or respond to posts with preprogrammed messages on specific issues.

An argument can be made for denying bots First Amendment protections. After all, to the extent that free speech is designed to facilitate and promote the democratic process, the process should protect the speech of human beings and not bots. But the dividing lines between bots and humans is not as clear-cut as one might think or hope. Bots rarely function completely autonomously. In some instances, bots simply provide the mechanism by which individuals widely distribute their own messages. In other instances, bots are programmed by humans

124 *Id.*
125 *See* MacFarquhar, *supra* note 119.

to respond on particular issues with pre-determined messages. In both instances, bots are subject to human programming and control. Even when a bot creates content, it's message is grounded in prior programming by a human. Thus, it seems inappropriate to simply disregard bot speech as machine speech rather than human speech.

As noted, bots can have an outsized influence on the public debate by flooding the internet (and social media) with their messages. However, it is difficult to deny the humans who operate these bots that opportunity. Media moguls have used machines (e.g., printing presses, radio and television) to exercise an outsized influence over the public debate for many centuries. Bots might be regarded as nothing more than the mechanism by which ordinary people obtain a comparable level of influence.

Of course, there is also the problem of disinformation. However, the U.S. system is not well-equipped to deal with that problem. Unlike other systems (e.g., the French Gayssot law), the U.S. Constitution does not allow the government to declare that certain ideas are sacroscant and cannot be controverted. Indeed, at the very foundation of the U.S. system is the idea that governmental orthodoxy is objectionable.

András Koltay, Budapest

# The Notion of the 'press' with Regard to the Freedom of the Press

## 1. Introduction

Freedom of the press is understood as a right that guarantees the freedom of all media. The press (as printed products or online news portals), television, radio and on-demand media services are subject to separate regulatory regimes and distinct constitutional principles in Europe. The freedom of online platforms that have characteristics other than those of legacy media but which play a similar role in the public sphere is treated as a separate issue. Freedom of the press is therefore the freedom of all channels of communication that can be covered by the notion of 'press' (with regard to freedom of the press) or 'media' generally. Of course, this definition in itself raises questions.

This analysis seeks to identify a European concept of the press that is acceptable and applicable to both theory and practice, by presenting the rules which exist in individual legal systems, the relevant case law of the European Court of Human Rights (ECtHR), while, with regard to the most important issues, I will also draw comparisons with the constitutional free speech doctrine of the United States.

## 2. The Concept of the Press

One of the fundamental questions regarding freedom of the press and freedom of expression is how to define the press, and whether such a definition is necessary at all. If we do not attribute independent content to freedom of the press, there is no need to define it. If we do, it quickly becomes clear that with the development of technology – and the spread of online communication – it is becoming less and less possible to define the press on the basis of its form or appearance, that is, based on technological characteristics. Yet in the past – as long as only the printed press, radio and television existed – this was the obvious and easy path. Today, producing such a definition is more arduous and more debatable, as it requires the identification of the tasks and social significance of the press.

## 2.1. Why is the Press Important?

Before attempting to clarify the concept of the press, it is worth examining why each constitutional system attaches great importance to the press. This issue is connected to the general justifications for freedom of expression. Democracy requires informed voters, and orientation to public affairs in general, even between elections, is necessary for the citizens of a democratic state, and this is an area where the press can play an important role. The search for the truth – even if truth identifiable by public discourse is seen as a kind of narrow-minded truth, a term used in a 'technical' sense that refers not to finding objective truth but to the way public disputes are resolved – can be pursued more effectively through the press. To this end, the activities of the press are subject to standards of conduct with regard to both the identification of important events and the collection, editing and dissemination of news, all of which increase the chances of the public finding the 'truth' – as well as equipping the community to make appropriate decisions.[1] If the press is thus regarded as a servant of democracy, the third major justification for freedom of speech, which encompasses the various individualist approaches, and seems to be the least relevant to the press. Although journalists have a right to self-fulfillment through their work, this aspect is already more difficult to apply to media owners and media companies.[2]

Freedom of the press is thus somehow linked to the concepts of public discourse and public opinion. In an ideal model of public discourse, the stakeholders talk to each other about the issues, problems and difficulties that arise in society and which affect small or large groups of people, they are informed about individual issues, all relevant opinions can be published and decisions can thus be made with the information required. Although this ideal is not necessarily realised in practice, the press aids society in its effort to achieve these objectives. This theory has been explained in detail in several works by Jürgen Habermas.[3] Public discourse and public opinion in a democracy involve open, free and debating people.[4] Hence freedom of the press also seems to have something to do with the

---

1 Jan Oster, 'Theory and Doctrine of "Media Freedom" as a Legal Concept', 5(1) *Journal of Media Law* (2013) 57, 68–72.

2 Ibid. 73.

3 Jürgen Habermas, *The Structural Transformation of the Public Sphere* (Cambridge, Polity, [1979] 2020); Jürgen Habermas, *The Theory of Communicative Action* (Boston, Mass, Beacon, [1981] 2005).

4 Jan Oster, *Media Freedom as a Fundamental Right* (Cambridge, Cambridge University Press, 2015) 29–30.

concept of democracy. According to a Council of Europe recommendation, the freedom of the media is 'indispensable for genuine democracy and democratic processes'.[5] Free media and democracy are commonly referred to as interdependent values,[6] while 'democracy' refers almost exclusively to liberal democracy by the authors who examine the theoretical issues of freedom of speech and press.[7] But can there only be true freedom of the press in a liberal democracy?

Several types of democracy are possible,[8] and it is worth recalling that liberalism, which places individual freedom as its core value, and democratic decision-making, which puts the will of the majority at the forefront, may work counter to each other, and liberal democracy must find a balance between these two values. In their work, which is fundamental to media theory, Daniel Hallin and Paolo Mancini also distinguish between 'liberal', 'democratic corporatist' and 'polarised pluralist' models within Europe, categorising the media system in each state,[9] while considering all the states they examine to be liberal democracies. What is more, as James Curran notes, this modelling – first and foremost the liberal model the British author is so interested in – 'is as much an elegant construct, as much as a work of creative intelligence, as the Hogwarts academy for young wizards.'[10] However, since this work is not concerned with the details of political theory, it is sufficient in this paper to interpret the term 'democracy' as encompassing all types of democracy, or at least all those that prioritize protecting freedom of speech and of the press.

Another question is whether there is such a thing as 'European freedom of the press', that is, whether there is a theoretical consideration, a constitutional approach or regulatory practice of freedom of the press on the continent that is found in all European democracies but which is different from other non-European states with distinct cultural traditions and history. Although the outlines, precise boundaries, doctrines and practices of freedom of the press – similarly to freedom of expression – may not be exactly the same in each European state, they have enough in common to allow an examination of 'European freedom

5 Recommendation CM/Rec(2011)7 of the Committee of Ministers to member states on a new notion of media, Section 2.

6 Karol Jakubowicz, *Media and Democracy in the 21st Century* (Warsaw, Poltext, 2013).

7 See, for instance, Perry Keller, *European and International Media Law* (Oxford, Oxford University Press, 2013).

8 Jacob Rowbottom, *Media Law* (Oxford, Hart, 2018) 22–24.

9 Daniel C Hallin and Paolo Mancini, *Comparing Media Systems* (Cambridge, Cambridge University Press, 2004).

10 James Curran, *Media and Democracy* (London, Routledge, 2011) 29.

of the press'. Jan Oster goes even further: He does not rule out creating a global, universal *jus commune* for freedom of the press,[11] although we are still a long way from it, despite the international human rights mechanisms working in this direction. Nevertheless, freedom of the press can, of course, include elements that apply equally or similarly to all democracies around the world.

Although the legal systems of all European states share many common elements in in terms of the scope of freedom of the press in the twenty-first century, and the underlying values that underpin it, the differences between them are also not insignificant. The attitudes of a country towards freedom of the press can only be understood in depth if its history is taken into consideration. The peculiarities of each country's path leading to the current state of freedom of the press explain the differences between their approaches. The histories of the UK,[12] other Western European countries[13] or individual post-communist countries, including of their struggles for freedom of the press, are fundamentally different, and that of the US all the more.[14] All of these historical differences also have important implications for freedom of the press, and it is inevitable that we will occasionally turn to the past. Although history has not ended with the fall of the Iron Curtain, despite hopes to the contrary,[15] the decades that have passed since have brought the political system and social life of the old and new European democracies closer together – not least as a result of the process of European unification, that is the strengthening of the European Union. Accordingly, the rules, practices and cultures of European states relating to freedom of the press are much more similar today than they were in 1989–1990, when the communist dictatorships collapsed.

11 Oster, *Media Freedom*, 10.

12 For a comprehensive history of English freedom of the press, see Paul Wragg, *A Free and Regulated Press* (Oxford, Hart, 2020) 10–25; James Curran and Jean Seaton, *Power Without Responsibility*, 8th ed. (Florence, Taylor & Francis, 2018).

13 Frede Castberg, *Freedom of Speech in the West* (London, Allen & Unwin, 1960).

14 Leonard W Levy, *Emergence of a Free Press* (Chicago, Ivan R Dee, 2004); Stephen M Feldman, *Free Expression and Democracy in America: A History* (Chicago, University of Chicago Press, 2015); David Anderson, 'Origins of the Free Press Clause', 30(3) *UCLA Law Review* (1983) 455; Sonja West, '"The Press", Then & Now', 77(1) *Ohio State Law Journal* (2016) 49–105.

15 Francis Fukuyama, *The End of History and the Last Man* (New York, Free Press, 1992).

## 2.2. Freedom of the Press and Freedom of Expression – Difficulties in Defining the 'press'

In order to identify the nature and limits of freedom of the press, it first seems necessary to clarify the concept of the press, and then distinguish between freedom of the press and freedom of expression. According to Jacob Rowbottom, the press can be defined in several ways, and definitions can be either 'institutional' or 'functional' in nature.[16] To this we can add an approach to the press as a 'technology', as Eugene Volokh argues that freedom of the press is nothing more than the right to freely use various technologies that facilitate mass communication, and thus the press is not privileged or burdened with additional obligations, and its boundaries coincide with those of freedom of expression.[17] However, as Owen Fiss eloquently demonstrates, the initial paradigm of free speech is the street corner speaker, in that legislators and, particularly in the US and the UK, courts have traditionally set limits on the restrictive powers of the state and government in order to protect individuals who are necessarily in a much weaker position than them.[18] In the age of media conglomerates, which grew enormously in the twentieth century,[19] and – in the twenty-first century – technology giants, which are many times larger in size and which dominate online communication,[20] the paradigm of street corner speakers may still be applied to these companies – to ensure their free, uninterrupted operation – but it clearly leads to grotesque results. Mark Zuckerberg, the powerful lord of Facebook, may invoke the same right as an unknown speaker at Speaker's Corner in London's Hyde Park if an authority or court wants to take action against him. Unlike street corner speakers, companies interested in public communication do not operate primarily to publicise their 'own' (their owners') opinions, but instead as businesses, yet even so their activities fundamentally affect the freedom of expression of others and, in general, the quality of the democratic public, as important actors of it.

---

16 Rowbottom, *Media Law*, 27–31.

17 Eugene Volokh, 'Freedom for the Press as an Industry, or for the Press as a Technology?', 160(2) *University of Pennsylvania Law Review* (2012) 459. For a summary of similar views of other authors, see Peter Coe, 'Redefining "Media" Using a "Media-as-a-Constitutional-Component" Concept', 37(1) *Legal Studies* (2017) 25, 37–39.

18 Owen M Fiss, 'Free Speech and Social Structure' 7 *Iowa Law Review* (1985) 1405.

19 Ben H Bagdikian, *The New Media Monopoly* (Boston, Mass, Meacon, 2007); Robert W McChesney, *Rich Media, Poor Democracy* (New York, The New Press, 2015).

20 Matthew Hindman, *The Internet Trap* (Princeton, Princeton University Press, 2018); James Ball, *The System* (London, Bloomsbury, 2020).

If the press is an important institution of the public, and thus of a democratic social order, then the question arises as to whether it is necessary for the legal system to provide the press with additional rights, or possibly impose additional obligations in return. If the press is an 'institution', this seems inevitable, but as Tom Gibbons notes, 'their association with truth and participation in a democracy is only incidental.'[21] Gibbons' tactful words highlight the fact that the press cannot be obliged to act in accordance with the justifications underlying the protection of freedom of expression, just as individuals cannot always be obliged to exercise their freedom of expression in the interests of democracy or truth, or possibly with a view to self-fulfillment, in accordance with one philosophical justification for freedom of expression. While no one expects it of the individual, we expect the institution of the press to serve democracy and contribute to the search for truth, at least in principle, but also as the desired outcome in legislation, court decisions, scholarly work, and public discourse. This expectation does not, however, create any legal obligation on any press product, yet it provides a basis for distinguishing freedom of the press from the exercise of freedom of expression.

It would be tempting to define the press in terms of its technological characteristics, and just a few decades ago this would not have raised any particular difficulties. Print newspapers, radio and television (these can be summed up as legacy media, as opposed to newer, mainly online channels of communication) – at most there may have been some controversy over the judgment of a freelance journalist. However, with the development of technology, previous frameworks have loosened, and public information is now provided not only by legacy media staff, but also by bloggers, vloggers, influencers, youtubers and, of course, any individual who has access to a public forum, typically online. It is not necessary for the speaker that provides information to be a professional journalist, nor is it necessary for him to carry out his activity on a for profit basis. The press can no longer be defined on the basis of purely formal criteria. It is worth determining whether it is reasonable to approach the press on the basis of its functions. Three functions can be identified that may bring us closer to establishing a useable definition of the press, and these are:

- informing the public about important and newsworthy events;
- monitoring the government as a public watchdog;
- promoting democratic dialogue, providing a platform for individual opinions.

21 Thomas Gibbons, *Regulating the Media* (London, Sweet & Maxwell, 1998) 28.

Providing information to the interested public is essential in a democracy, and is a key element of the press's self-image and self-reinforced image. Sonja West considers this, along with supervising the government as unique constitutional functions of the press.[22] The press is a symbolic agora of today's society;[23] it is the space where individual opinions may appear and clash, and which also helps those interested in making informed opinions and making decisions. Individuals need the help of the press without which they would not have access to necessary information. It would be a mistake to idealise the relationship between the press and its audience, as the former does not necessarily provide information properly and the latter does not necessarily take an interest in public affairs information.[24] The former problem can be dealt with in part through legal regulation, while the latter has nothing to do with press regulation.

The idea of the press keeping watch over the government can be linked to a metaphor in Thomas Carlyle's Fourth Estate.[25] According to this, the press exercises control over the current power, revealing the functioning of state bodies and government, exposing its anomalies and providing the members of society with sufficient information in order to be able to make informed decisions in democratic procedures. It should be noted, however, that Jeremy Bentham had already recognised this before Carlyle[26] and James Mill[27] also highlighted the supervisory role of the press. This function essentially merges with the general informative role of the press, yet it is not unreasonable to highlight it, nor is it a coincidence that some authors highlight the strong suspicion of government in

---

22 Sonja West, 'Stealth Press Clause', 48 *Georgia Law Review* (2014) 729, 749–755; see also Sonja West, 'Awakening the Press Clause', 58 *UCLA Law Review* (2010–2011) 1025, 1068–1070; Sonja West, 'Press Exceptionalism', 127(8) *Harvard Law Review* (2014), 2443.

23 Thomas G Tucker, *Life In Ancient Athens: The Social and Public Life of a Classical Athenian from Day to Day* (London, Macmillan, 1906).

24 Rowbottom, *Media Law* 18–19.

25 Thomas Carlyle, *On Heroes and Hero-Worship and the Heroic in History* (London, James Fraser, 1841).

26 Jeremy Bentham, 'Letter to the Spanish People I on the Liberty of the Press, the Approaching Eight Months' Sleep of the Cortes, and the Exclusion of Experience from the Succeeding Cortes' in Jeremy Bentham, *On the Liberty of the Press and Public Discussion* (London, William Hone, 1821) 9, 12.

27 James Mill, 'Liberty of the Press' in *Encyclopaedia Britannica* (Edinburgh, Adam and Charles Black, 7th edn, 1842) vol VIII, 272, 278.

the US, which has been a basic experience since the beginning of its history.[28] Nevertheless, the idea itself comes from Europe, where government control of the press is identified as an important function of the press,[29] and the ECtHR often refers to the press as the public watchdog if it wants to justify extending its freedom.[30]

The third essential function of the press is to promote democratic dialogue by providing a channel for different opinions. This idea was formulated long before the emergence of democracies in today's sense.[31] It can also mean that journalists report on individual positions themselves and that they allow room for other speakers. It is important that the press does not transmit individual opinions to the audience 'raw' without sorting and processing them, but arranges them, puts them in context, and confronts them with other opinions. The press is the principal organiser of the public dialogue.[32] However, it should also be noted that tensions may arise between the individual wishing to express his views and the press, due to the latter's decision-making role as an organizer.[33]

After reviewing the most important functions of the press, it is clear that simply listing them is not sufficient to distinguish the press from other public speakers who are also capable of performing one or all of these functions. These may be individual speakers, bloggers or social media users, but they can also be

---

28 Vincent Blasi, 'The Checking Value in First Amendment Theory', 2(3) *American Bar Foundation Research Journal* (1977) 521; Potter Stewart, '"Or of the Press"', 26(3) 26 *Hastings Law Journal* (1975) 631.

29 Rowbottom, *Media Law*, 19–20.

30 *Lingens v Austria* no 9815/82, judgment of 8 July 1986, para 44; *Jersild v Denmark* no 15890/89, judgment of 23 September 1994 [GC], paras 31 and 35; *Bladet Tromsø and Stensaas v Norway* no 21980/93, judgment of 20 May 1999, paras 59 and 68; *Observer and Guardian v the United Kingdom* no 13585/88, judgment of 26 November 1991, para 59b; *Sunday Times v the United Kingdom* no 13166/87, judgment of 26 November 1991, para 50b; *Dalban v Romania* no 28114/95, judgment of 28 September 1999, para 49; *Bergens Tidende and Others v Norway* no 26132/95, judgment of 8 August 2000, paras 49 and 57; *Thoma v Luxembourg* no 38432/97, judgment of 29 March 2001, para 45; *Goodwin v the United Kingdom* no 28957/95, judgment of 27 March 1996, para 39.

31 Daniel Defoe, *An Essay on the Regulation of the Press* (London, 1704); Mill, 'Liberty of the Press'.

32 David Anderson, 'The Press and Democratic Dialogue', 127 *Harvard Law Review Forum* (2013–2014) 331, 333. See also David Anderson, 'The Press and Political Community' in András Koltay (ed.), *Comparative Perspectives on the Fundamental Freedom of Expression* (Budapest, Wolters Kluwer, 2015) 41–51.

33 Rowbottom, *Media Law* 22.

the service providers themselves, such as a blogger, video sharing, or social media platform. It is welcome if, for example, a court grants an individual speaker who is engaged in information activities strong protection similar to that of the professional press,[34] but it raises the question of whether it is necessary to grant him privileges as well, and whether such a speaker has any responsibility to the public.

In order to clarify the concept of the press, it is not enough to define its functions; it is also necessary to identify other distinguishing features. Some of these can be found – all of which will be features that an individual speaker or, for example, an online platform may have, but which are entirely specific to the press, and presumably only a part of it. One of the most obvious such features is mass communication, which is undoubtedly one of the peculiarities of the press: Its purpose and life force is to address many people.[35] The professional nature of the operation, and the knowledge of and adherence to standards of conduct specific to the activity is another distinguishing feature.[36]

The Recommendation of the Council of Europe referenced above also called for a new definition of the media.[37] Its starting point is that the 'new ecosystem' of the media includes all the new actors who take part in the process of content production and distribution, transmitting content to a potentially large number of people (content aggregators, application designers, content-producing users, and businesses which ensure the operation of the infrastructure), as long as they have some editorial influence on the content or supervise it in some way.[38] Hence, the defining element is the existence of editorial responsibility. The Appendix to the Recommendation sets out six criteria. If all of them are met, a service can be regarded as media:

(1) the intent to act as media,
(2) the purpose and underlying objectives of media – produce, aggregate or disseminate media content,
(3) editorial control,
(4) professional standards,
(5) outreach and dissemination,

---

34 Jan Oster gives examples of this from the practice of the ECtHR, see Jan Oster, *European and International Media Law* (Cambridge, Cambridge University Press, 2016) 11–12.
35 Oster, *Media Freedom*, 60–61; Coe, 'Redefining "Media"', 39–40.
36 Coe, 'Redefining "Media"', 40–41.
37 Recommendation CM/Rec(2011)7 (n 7).
38 Ibid., paras 6–7.

(6) compliance with public expectations, such as accessibility, diversity, reliability, transparency, etc.

The Recommendation of the Council of Europe does not provide a precise concept, just certain factors that may be helpful for defining the press. Its weakness is that the individual aspects themselves give rise to a large number of disputable questions of detail (what are standards of conduct and who sets them; what are the expectations of the public and who identifies them, etc.). West highlights external recognition as press, self-identification as press, training as a journalist, participation in education, experience and regularity of publication, and a permanent audience as criteria for belonging to the press.[39]

It is important to keep in mind that defining the concept of the press, one which is acceptable in principle and workable in practice, is important only in so far as legislation distinguishes between the press and other public speakers; if the press enjoys special privileges access to those rights must be restricted so that the rights can be used in practice for the purposes established in principle. Therefore, in an age of mass public speech it may be necessary to identify which speakers qualify as the 'press', based on the nature of the information they publish.

David Anderson suggests that a decision should be made, based on the principles of the freedom of the press, as to which information is more important. This idea implies that only a speaker who, in terms of the privileges granted to the press, performs the task of informing about public affairs could be considered as "the press."[40] Jan Oster argues for a similar approach: For the purposes of regulation, only a 'natural or legal person gathering and disseminating to a mass audience information and ideas pertaining to matters of public interest on a periodical basis and according to certain standards of conduct governing the newsgathering and editorial process' could be regarded as media.[41] Peter Coe suggests the following criteria for defining the concept of media:

- a natural and legal person,
- engaged in the process of gathering information of public concern, interest and significance,
- with the intention, and for the purpose, of disseminating this information to a section of the public on a regular basis,

39 West, 'Press Exceptionalism', 2456–2462.

40 David Anderson, 'Freedom of the Press', 80(3) *Texas Law Review* (2002) 429, 512–515 and 528–530.

41 Oster, 'Theory and Doctrine', 74–77.

- while complying with objective standards governing the research, newsgathering and editorial process. These standards would include, for instance, the time spent researching stories and ensuring the provenance and reliability of information.[42]

However, the link between the concept of the press and the activity of providing information in the public interest has not been made in the legislation, and this is not easy to imagine in the near future.

In this examination of the conceptual elements, I have left perhaps the most difficult aspect until the last: the independence of the press. The ideal of freedom of the press is that the press is 'independent'. This requirement entails independence from the government or from state bodies,[43] which does not mean that a newspaper cannot agree with a decision of the government or support the government on a principled basis as long as it does so without coercion. 'Independence' also includes independence from the advertisers and sponsors who provide the economic basis for the media. According to some approaches, the journalist should be independent even of the owner of his newspaper, and the will of the owner should not directly influence the content of the newspaper. However, it would be insensitive to reality to fix the independence of the press as a conceptual element of the press because, as James Curran and Jean Seaton explore in their comprehensive work on the history of the British press, it is in fact a myth.[44] This does not undermine the merits of journalists and press outlets who really succeed in achieving their own independence, but the vulnerability of the media to market forces, the economic or political interests of owners and the aspirations of governments and other political forces to shape public opinion all cast doubt on the idea of independent journalism.

## 2.3. Press and Co

When discussing freedom of the press, we must also pay attention to other services of the legacy media beyond the press – television, radio and on-demand media services – while examining the regulation of the press only with regard to printed and Internet press products. With regard to freedom of the press, moreover, in the case of speakers other than those of the legacy media – freelance

42 Coe, 'Redefining "Media"', 51.

43 As Randall Bezanson emphasizes, Randall Bezason, 'Whither Freedom of the Press?', 97(4) *Iowa Law Review* (2012) 1259.

44 Curran and Seaton, *Power*, see in particular: 'Part I: Press History'.

journalists, bloggers and a host of new characters in online communication – we need to examine whether the public communication they perform can be considered an exercise of freedom of the press as has already been touched on in the previous paragraph in search for the definition of 'press'.

However, while television, radio and, for some time, on-demand media services should of course be regarded as practitioners of freedom of the press, we blithely deny this status to books, movies or individual speakers who use the telephone and mail as a means of communication. In the case of books, the general rules on freedom of expression prevail, providing adequate protection for the author, although after the invention of printing, the Catholic Church made the first provisions on 'media regulation' for books and institutionalised censorship, then compiled a list of prohibited books.[45] For films to be shown in cinemas and later released on a storage medium independent regulation applies, for example, for child protection purposes,[46] and their publication constitutes an exercise of freedom of the press when it takes place on television or any other audiovisual media service. For those communicating by telephone, post or e-mail, the mass nature of communication is lacking, although it should be noted that both the press and the media can use e-mail (newsletters) and mail (delivering printed mail to the public).

The 'press' (and the media) means both the provision of content and the use of technology to deliver it to the public, as well as the use of certain intermediary services in the delivery process.[47] The concept of freedom of the press is somewhat broader because both criteria are essential for the exercise of this right. The activities of intermediary service providers (postal services in the offline world, courier services, Internet service providers on the Internet, social media platforms, video sharing sites, news aggregators, search engines, etc.) are essential to the exercise of freedom of the press, but do not constitute it in themselves, because the content displayed by them is not produced by these intermediaries or commisioned by them, that is the service they provide does not involve any editorial decision-making.[48] These intermediary services do not make decisions about the content of the press, but only transmit it to the audience and users.

---

45 Max Lenard, 'On the Origin, Development and Demise of the Index Librorum Prohibitorum', 3(4) *Journal of Access Services* (2006) 51.

46 For example, the film industry in the UK is regulated by the British Board of Film Classification which is responsible for classifying films that have a cinema and DVD release.

47 Oster, *European and International Media Law*, 3–5.

48 Ibid 13–15; Oster, *Media Freedom*, 58–60.

However, these services are also capable of interfering with content, which is similar to traditional editing in terms of the outcome of their activities, but the nature of the decisions they take is fundamentally different.[49]

## 3. Who has the Right to Freedom of the Press?

The question of who holds the right to freedom of the press is closely linked to the complications surrounding the concept of the press. Knowing who is interested in exercising freedom of the press, brings us closer to identifying those who are entitled to it. The more we approach the concept of the press from the point of view of the interests of the public and of society, the more complex the issues we face in defining the scope of beneficiaries. The problem of the freedom of the journalist vis-à-vis the owner of the press – his own employer – arises and it is also necessary to address the hypothetically existing, that is imaginary 'public' in exercising freedom of the press in the public interest. We call the public imaginary because, although it does exist by definition, freedom of the press typically identifies the public not as a community of actual rightholders or stakeholders, but as an abstract being whose needs, such as public interest content, are supposed to be identified by legal regulation on the basis of considerations of principle.

### 3.1. Journalists, Editors and Owners

At first glance, freedom of the press can be seen as the right of journalists and editors to determine the content of a given medium. However, attempting to define the category of 'journalist', raises similar difficulties as those encountered when trying to define the 'press'. A journalist is traditionally a professionally trained speaker who makes a living from this profession, as a professional employee in a particular medium or media. However, reaching out to a wide range of people (mass communication) is no longer the prerogative solely of legacy media, and these days traditional journalistic activities can also be carried out by other, untrained, non-professional speakers.[50] The press and professional journalists would naturally assert that there is a significant difference between the level of activity of professional and amateur journalists, and hence that strengthening the rights of the former is also in the interest of society. This viewpoint is also supported in some cases by legislation and case law, where certain

49 See András Koltay, *New Media and Freedom of Expression* (Oxford, Hart, 2019).
50 Oster, *Media Freedom*, 61–64.

privileges granted to the press are reserved only for certain legacy media services or their journalists.[51] However, this formal approach puts freelance journalists at a disadvantage, as well as all those who, although not journalists in the traditional sense of the word, contribute to informing the public, to the discussion of public matters.

Oster suggests that a speaker who reports on the discussion of public affairs is to be considered a journalist, irrespective of where he works, and this would be in line with the interpretation of the concept of the press which holds that the press includes forums reporting on public affairs.[52] A similar problem arises here: The press cannot be forced to report on public affairs, but excluding media that ignore public affairs from exercising freedom of the press would inevitably curtail that freedom and require problematic demarcation decisions from the courts. In practice, however, the issue does not arise in general terms, but in the context of the exercise of the additional rights (privileges) granted to the press and the liability – to a much lesser extent – arising from the additional obligations imposed on the press. In other words, the question in general is not who a journalist is, but whether a blogger can exercise the right to protect her source, or whether the journalists of a tabloid newspaper also enjoy that right. As we will see later, these rights and obligations will be able to be formulated in regulations and court decisions in such a way that they meet both the general aspect of freedom of the press and the specific requirement to discuss public affairs. The Irish High Court's decision in *Cornec v Morrice*[53] is a good example of this: According to this judgment, a blogger involved in public discourse may also claim the right to protect journalistic sources.[54]

As Edwin Baker argues, it would be naive to think that freedom of the press belongs exclusively to journalists, however we define them.[55] The owners of the press, or even company executives acting on behalf of the owners, can also invoke freedom of the press against external restrictions. Internal power relationships, ownership and freedom of the press can also come into conflict with each other,

---

51 See, for example, *Perrin v the United Kingdom* no 5446/0318, judgment of 18 October 2005.

52 Oster, *Media Freedom*, 65–66.

53 *Cornec v Morrice* [2012] 1 IR 804.

54 See Eoin Carolan, 'Promoting Civic Discourse: A Form of Positive Free Speech under the Constitution of Ireland?' in Andrew T Kenyon and Andrew Scott (eds), *Positive Free Speech: Rationales, Methods and Implications* (Oxford, Hart, 2020) 65, 72.

55 C Edwin Baker, *Human Liberty and Freedom of Speech* (Oxford, Oxford University Press 1989) 253–255.

a conflict that is difficult to resolve without compromising the fundamental considerations behind freedom of the press. If an owner dictates to a journalist what to write about or determines how he or she can do so, it is a restriction on the freedom of the journalist, but does not necessarily infringe on the freedom of the press, which the owner may rightly invoke. Although concerns typically revolve around the freedom of the journalist and the fear that the owner may be able to influence public information for economic or political reasons, it must be recalled that the press is fundamentally an economic enterprise and the owner is responsible for its proper functioning, putting his own financial well-being at risk. As will be seen below, legal systems rarely attempt to resolve potential conflicts between the interest of public information and the operation of the media. It follows that the freedom of the owner is, in practice, stronger than that of the journalist, and that owners have the last word with regards to the content published and, ultimately, the continued employment of the journalist.

## 3.2. Individual and Institutional Speakers

The difference between 'individual' and 'institutional' speakers stems from a similar root as the difference between the freedom of the journalist and the owner. Freedom of the press is an 'institutional right', in that it belongs not only to the individual (journalist and editor) but also to certain organisations and enterprises of the press, that is the institution.[56] It can be argued that the press as an institution has less grounds for freedom of expression than an individual who expresses his views publicly, given that the press does not publish the personal opinions of the journalists working there but instead content is created through the work of many people, as well as being influenced by ownership decisions and external influences. The press is under economic – and often political – pressure, as the interest of its owners is to gain influence and profit, while participation in public discourses is only secondary, so freedom of the press must be interpreted in the light of this reality. However, as Rowbottom explains, this is a questionable approach. Despite its institutional peculiarities, the press is made up of individuals whose opinions appear in the public sphere, filtered through the institutional framework.[57] While economic considerations do influence the content of the press, on the one hand, the pursuit of profit is a legitimate, appreciable interest, and on the other hand, economic success is the basis of the operation

56 Frederick Schauer, 'Towards an Institutional First Amendment', 89 *Minnesota Law Review* (2005) 1256.

57 Rowbottom, *Media Law*, 8.

of the press, without which the press could not exist. The profitable operation of the press is a necessary condition for the press to operate at all, and thus for the public interest to be taken into account in the public sphere, since if there were no market-functioning press, we would not be able to talk about its public interest obligations. However, all this cannot justify a broad interpretation of freedom of the press which does not take into account the damage which may be caused by the exercise of that freedom.[58] Furthermore, it would be a mistake to pretend that economic or other motivations of public speakers are specific to the press, and the individual speaker is necessarily freer than the institutional press. An individual's speech may similarly be influenced by social pressure, professional interests, and many other circumstances, limiting his 'free' expression.[59]

Clearly some confusion surrounds the ideal of 'independence'. On the one hand, we are voicing the ethos of the independent press, seeing it as a fundamental supporter of democracy, while at the same time we realise that the press cannot be completely independent for a number of reasons, and journalists cannot even be independent of their own superiors or the owner of their employer. On the other hand, the rights of an individual – who may find it easier to preserve his independence, at least in principle – who is potentially more 'independent' than the institutional press are defined by legal systems in a narrower way or with a different content compared to the press. This is only an apparent contradiction, however, as despite all the difficulties facing the press as an institution and the limitations of its operation, it is naturally more suitable for effective participation in the public than the less audible individual. As Curran puts it, 'The development of journalistic professionalism promoted, they tell us, media independence not merely from government but also from "parties and other political actors".'[60]

## 3.3 Freedom of Distributors and Intermediaries

Service providers whose activities are necessary for certain products of the press to reach the public may also claim protection by freedom of the press. If their activities are subject to a restriction on the content and services they can transmit or broadcast, this may be construed as a restriction on freedom of the press, even if the activities of those providers do not in themselves constitute the exercise of freedom of the press. In the case of the print media, this may apply to

58 Gavin Phillipson, 'Leveson, the Public Interest and Press Freedom', 5(2) *Journal of Media Law* (2013) 220, 232–234.
59 Rowbottom, *Media Law*, 10.
60 Curran, *Media and Democracy*, 31.

postal services and distribution in street and public shops, while for media services, it may be the broadcaster, and in the case of online content services, various intermediary service providers, in particular, Internet service providers.[61]

## 3.4 The Interests of the Audience

The concept of freedom of the press, which focuses on the open debate on public matters, considers the constitutional protection of the press to be justified in the public interest or, more narrowly, in the interest of the public. Accordingly, while freedom of the press is arguably a right enjoyed by the press, the regulation on the freedom of the press must take the interests of the audience into account.[62] According to this view, the primary interest of the public is to have access to information on public affairs and the widest possible choice between individual speakers and dissenting opinions in the press market. This kind of 'freedom for the public' is uncertain in content, however, and certainly does not create a direct right of access to certain areas of the press for members of the public (except in very narrow and exceptional areas such as the right of reply).[63] The interests of the public may be served by the press, but it does not have an enforceable right vis-à-vis the press or a third party involved in the communication process, in the context of freedom of the press.

Protecting the interests of the public is an indispensable basis for the freedom of the press, highlighting its role as a community watchdog and as the Fourth Estate, but this foundation becomes shaky when we admit that nothing requires the press to provide information in the public interest. As with the previous argument, this means that the public interest is only an incidental reason for certain additional rights to be specifically granted to the press dealing with public matters. Furthermore, Paul Wragg warns that citing the public interest is not without dangers. The press dealing with public affairs assumes that the citizens are interested in public matters. As such, the public interest is nothing more than the hope that such responsible citizens exist and demand information on public affairs. According to Wragg, prioritising this task for the press ultimately indicates to the audience what important information is worth knowing. This, in turn, runs counter to press pluralism if it seeks to favour a press that provides one type of information over other press outlets that do not harbour

61 Oster, *Media Freedom*, 81–82.

62 Andrew T Kenyon, 'Complicating Freedom: Investigating Positive Free Speech' in Kenyon and Scott, *Positive Free Speech*, 7.

63 Baker, *Human Liberty*, 251–253.

ambitions of presenting public affairs.[64] The courts sometimes indicate in their decisions which information it is in the public interest to be published and make it clear that this is not the same as information that is otherwise of interest to the public.[65] As long as the aim is not to protect the freedom of the press as a whole, but only certain privileges granted to the press in the public interest, more precisely in the interest of publishing content on public affairs, this should not be such a major concern. Similarly, an approach which disproportionately restricts the freedom of the press to the detriment of public life must ensure that the legal system provides greater protection for the openness of public discourses, for example as regards the limits to the protection of privacy, than for the freedom of the tabloids.

## 4. Conclusions

Various difficulties arise when defining the concept of the press. It is not appropriate to consider the use of certain technologies as a decisive conceptual element, given the wide availability of some of these means. It may seem more promising to start from the functions and tasks of the press, but highlighting certain activities performed by the press, such as providing the public with information on public issues, excludes media practitioners that do not bother with this task from being recognised as belonging to the press, as well as opening the concept to individual speakers. It would seem that additional distinctive features are needed, and chief among these are professional operation, recognition and observance of professional norms.

Clearly defining the concept of the press is of limited significance only if the legal system grants or imposes additional rights – and possibly additional duties – on the press. The identification of the press is therefore necessary in relation to situations where its regulation distinguishes between the press and other speakers. This happens in all legal systems, but most of the rules for public speaking are universal, covering everyone.

---

64 Wragg, *A Free and Regulated Press,* 241–251.

65 This is especially common in cases involving the boundaries of celebrity privacy, compare, for example *Von Hannover v Germany* no 59320/00, judgment of 24 June 2004; *Tammer v Estonia* no 41205/98, judgment of 6 February 2001. However, if information about a celebrity is in the public interest, it may be protected by freedom of the press, *Axel Springer AG v Austria* no 39954/08, judgment of 7 February 2012, [GC]; *Von Hannover (No 2)* no 40660/08 and 60641/08, judgment of 7 February 2012 [GC], and *Von Hannover (No 3)* no 8772/10, judgment of 19 September 2013.

The right to freedom of the press belongs to all stakeholders involved in the communication process, and these rights can conflict with each other (for example, between a journalist and a press proprietor). At the same time, the public or, more generally, society may have no legally recognised rights, only certain interests, which are taken into account by press regulation and relevant case law.

Michael Losavio, Louisville

# Freedom of Speech, Inviolability and Democratic Cyber Security – an End or a Beginning?

## What Happens When Everyone Knows Everything About Us?

## I. Introduction

The electronic frontier has brought us to new opportunities for the exchange of ideas, free expression and free association. This frontier is one of information and knowledge. But it offers opportunities for suppression, abuse and criminality, especially through attacks by people against the ideas of those with whom they disagree. As on the frontiers and marchlands of history, law and lawlessness wage ongoing battles for justice and injustice. And, as in history, the borders of those frontiers may be ill-defined, leading to more conflict through assertions of power and jurisdiction.

Historical equivalents suggest means to protect free speech and its speakers in this modern technological era, where the inviolability of speech and speakers is guaranteed less and less in both autocracies and democracies. The metaphorical Horsemen of the Infocalypse ride as both public and private threats. This seems anomalous in these contemporary, modern rule-of law days, but this highlights the gaps in the rule of law globally and the regulation of technology generally. The electronic frontier is the entire world, as Information and Communication Technologies (ICT), especially the Internet and cellular telephone networks, span nations.

That globe-spanning information exchange runs through multiple national jurisdictions, many of whose law, policies and agendas conflict even with their neighbors next door. Functionally everyone is right next door in the topography of the Internet, whether local police or authoritarian state security on the other side of the world. But this may be a one-way function, as the attacker may be anonymized or obfuscated as to make accountability difficult or impossible.

The exchange of ideas must consider this and the risks engendered. If we must default to historical models of self-protection, in ICT the discipline of cyber security must be distributed and democratized to give that protection to speech and speaker.

## II. The Risks to Security and Free Expression

These risks to public and private security and private rights of free expression are many, with new malfeasants engaging in malicious activities in ICT. Contemporary threats on this frontier are from:

1. State actors and public security,
2. State-sponsored actors, and
3. Private parties, acting out of malice, hatred or profit.

System capabilities of response to these are inadequate. In modern democracies freedom of speech assumes a level of protection for the speaker, regardless of their risks and courage in speaking. That protection ranges from bars to state punishment and interference of speech to state protection from private retribution for speech. It is part of what Yemini calls the "inviolability" for speech, which he deems threatened by engineering systems of current use, global systems speech through our information and communication technologies. If the Internet is to remain greatest engine of freedom of expression and access to knowledge in human history, as the US Supreme Court noted in *ACLU v. Reno*, engineering of these technologies must include means by which to protect the inviolability of the speaker.

This inability to protect speech, much less any cyber systems from attack, has cost money and suffering. This failure and its concomitant frustration have led to suggestions for resort to the provisions under the Constitution (US) to issue letters of *marque and reprisal* to entities for cyber attacks.[1], [2] This near-piratical regime offers immunity and authority for private persons and organizations to attack and counterattack malicious cyber attacks wherever in the world the source of these attacks are found. In itself such a suggestion acknowledges the failure of organized public protection, and the last resort of vigilantism to protect the public.

---

1 Thomas Ayres, "A Maritime Solution for Cyber Piracy," Wall Street Journal, 13 May 2021, https://www.wsj.com/articles/a-maritime-solution-for-cyber-piracy-11620922458

2 Still, Jonathan "Resurrecting Letters of Marque and Reprisal to Address Modern Threats," Master's Project, US Army War College, March, 2012, https://apps.dtic.mil/sti/pdfs/ADA590294.pdf accessed 20 March 2021.

But there are other means private citizens have used within the rule of law to protect themselves in the physical world. That private guardian status can be extended to the protection of free expression in the online world.

Those protections are contained in the principles and systems of cyber security. But those same principles and systems may be a threat, as well. We discuss the balancing of actions and threats in technology and law.

## III. ICT and the US Constitution's Protection of Speech and Association

The U.S. Supreme Court in its modern jurisprudence on Information and Communication Technology systems suggests a framework for resolving constitutional issues with it; this framework balances the impact of ICT on liberties against the benefit to society in general and law enforcement in particular.[3] This reasoning can be seen in the 4th Amendment case of *United States v. Carpenter*, where the Supreme Court modified doctrine as to the reasonable expectation of privacy in information held by third parties, and found as to historical cell site data that an individual did indeed have such an expectation of privacy from the unprecedented intrusiveness of those systems, requiring a warrant for state access. For freedom of expression, ICT "... provides more expressive capacity to individuals than ever before, also systematically diminishes their liberty to speak." due to an unprecedented lack of anonymity and lack of inviolability.[4] One aspect of this tension is cyber security. A wide ranging domain, systems of cyber security can protect expression by offering confidentiality, integrity and availability of the means of expression.

Yet the increasing deployment of cyber systems and of cyber security threaten these protections. A violation of privacy may be just a click away, whether from state, state-sponsored or private actors. Establishing a balance, in technology and in law, will be an ongoing challenge for all of us.

---

3 Bhagwat, et al. note the Supreme Court of the United States has set out a template for resolving this tension between technological "Candides" (optimists) and "Cassandras" (pessimists) as to the impact of these technologies. He does not know if the Candides will support over time unregulated technologies as the impact of use and misuse evolves. Not does he know if the Cassandras will limit regulation to current "compelling needs" or expand as to interference in democratic processes.

4 Yemini, Moran, "The New Irony of Free Speech," 20 Columbia Science and Technology Law Review 119 (2018).

## IV. Case Study of a Violation of Autonomy to Speak- Sony Entertainment & The Democratic Peoples Republic of Korea

Wilton analyzes one very public example of this tension, especially as to inviolability, in the 2014 cyberattack against Sony Pictures by North Korean actors retaliating for its production of the film "The Interview" parodying the leader of North Korea.[5] Compromising about 38 million company files, including employee emails and embarrassing communications about Hollywood figures, this led Sony to censor the film. Further, it may have led others in the entertainment industry to censor their projects out of caution, undermining their freedom of expression and that of people to receive information.

Bruni posited an even wide impact, with the very high profile hack of Sony Entertainment roiling high-profile relationships throughout Hollywood leading to public recognition that confidentiality of electronic information cannot be guaranteed.[6] It is a warning to every single person speaking through ICT, often on very personal or very political matters that may impact their lives.

Yet as Wilton notes, even awareness of these risks does not deter people from using digital systems.[7] Citing Brandeis, Bentham and Orwell, he notes that the possibility of surveillance will affect how people act. But the nature of that, and the willingness and means of people to change or limit their behavior in email, texts or online social media is unknown. Certainly anonymous communications sought by criminals indicates that proper motivation encourages cyber secure systems.[8]

---

5 Conrad Wilton, Sony, Cyber Security, and Free Speech: Preserving the First Amendment in the Modern World, 7 Pace Intell. Prop. Sports & Ent. L.F. 1 (2017) (reprinted from 16 U.C. Davis Bus. L.J. 309 (2016)), http://digitalcommons.pace.edu/pipself/vol7/iss1/1

6 Frank Bruni, Hacking Our Humanity, N.Y. TIMES (Dec. 20, 2014, 9:53 PM), http://www.nytimes.com/2014/12/21/opinion/sunday/frank-bruni-sony-security-and-the-end-ofprivacy.html.

7 Id, at p. 17.

8 Certainly the interests of criminals in anonymous speech changes their behavior, as the recent US-Australian sting using the supposedly secure Anom app indicated; unfortunately for them, Anom was controlled by the FBI and Australian federal police, and was used to develop evidence for hundreds of arrests across the world. See Packham, Colin, "Global crackdown on organized crime after high-tech U.S.-Australia sting." Reuters Asia Pacific, https://www.reuters.com/world/asia-pacific/australian-police-arrest-over-200-after-cracking-underworld-messaging-app-2021-06-08/ 8 June 2021, accessed 8 June 2021.

This reflects the impact of pervasive, ubiquitous computing. From cell phones to Edge computing to Cloud computing, each person generates a massive amount of data about their speech, their locations at particular times, and their associations. From that data the great power of data analytics and artificial intelligence (AI) technologies permits direct knowledge and powerful inferences of what a person is doing. This is mirrored by the growth in suitable targets for cyber attacks against the security of everyone in myriad ways. New concerns come from new information technologies, their constructs and how we secure the constellation of data generating, transmission and collection, especially with the:

1. Internet of Things
2. Big Data
3. Cloud, Fog and Edge computing
4. AI and machine learning

This interplay of data and analytics with systems of security leads to vast direct and inferential evidence of a person's life, particularly their thoughts, speech and associations.[9] Examples of speech, conduct and association, direct and inferential, include:

1. Speech and statements,
2. Place, location and Geo-tagging,
3. Time and history,
4. Associations with others,
5. Association with activities.
6. Association with particular beliefs.

The right of personal autonomy, sometimes styled privacy, and free speech are intertwined, with American constitutional protections for protections of disclosure of associations and anonymous speech, even if offensive to some. Thus "Anonymity is a shield from the tyranny of the majority."[10] The Supreme Court reasoned "It thus exemplifies the purpose behind the Bill of Rights, and of the First Amendment in particular: to protect unpopular individuals from, retaliation-- and their ideas from suppression--at the hand of an intolerant society." Similarly, the Supreme Court observed freedom of association essential of the "liberty" assured by the Due Process Clause of the Fourteenth Amendment; the " …

9 Losavio, M. (2019) "Pensees': Speech, Inference and the Information Panopticon," R. Weaver, M. Cole and S. Friedland, (Ed.) *Comparative Perspectives on Freedom of Expression*, CRC Press.

10 *McIntyre v. Ohio Elections Commission*, 514 U.S. 334 (1995).

compelled disclosure of petitioner's membership lists is likely to constitute an effective restraint on its members' freedom of association".[11]

## V. Democratic Cyber Security and Freedoms of Expression and Association

The cyber security of these systems is critical to protecting these rights of free expression as well as privacy of the individual. Further, artificial Intelligence and its many diverse manifestations raise additional issues regarding those rights through inferences and pattern recognition previously difficult or impossible. These, in turn, push cyber security to the forefront of our concerns, just as physical violence and its prevention directs general police action. A key aspect of what we must prepare is an age old conflict brought forward: how do we protect from attacks against our rights and liberties while not simultaneously destroying those rights and liberties?

Noting this concern, Chemerinsky wrote "[I]t is so important for the debate to get past the point where one side is saying, 'We've got to give up civil liberties,' and the other side is saying, 'We cannot give up civil liberties' … It has to be a much more nuanced discussion of what civil liberties are being compromised, under which circumstances, and for what gain."[12]

The democratization of technology and the democratization of knowledge are sometimes used to address the process by with both become more available to more and more people in society. This ranges from the open source movement to ease of use of proprietary devices. It supports inclusion, opportunity and accountability.[13, 14, 15]

Law enforcement and public safety under the democratic accountability implicit in the American Constitution are vital to the liberties and protections

11 *NAACP v. Alabama ex. Rel. Patterson*, 357 U.S. 449 (1958).

12 Arbus, Melissa. "A Legal U-Turn: The Rehnquist Court Changes Direction and Steers Back to the Privacy Norms of the Warren Era." *Virginia Law Review* 89, no. 7 (2003): 1729–777. Accessed June 8, 2021. doi:10.2307/3202362.

13 Veak, T.J. (2006). Democratizing Technology: Andrew Feenberg's Critical Theory of Technology. NewYork: SUNY Press.

14 Feenberg, A. (1999). Questioning Technology. Routledge. ISBN 0415197546.

15 Veak, T. (2000). "Whose Technology? Whose Modernity? Questioning Feenberg's QuestioningTechnology." *Science, Technology and Human Values* 25, no. 2: 226–237. JSTOR 690112. (https://www.jstor.org/stable/690112).

of citizens.[16] A subset is the concept of community policing, a close engagement between law enforcement and the people to protect.[17] Together they may assure better community protection and engagement under the rule of law. Each alone is insufficient.

Why does cybersecurity matter? Consider these aspects of cybersecurity protecting freedom of expression and association:

| **Protections of free expression** | **Cybersecurity protection** | **Cybersecurity threat** |
|---|---|---|
| Media of communication via the Internet | Protects interference and modification of the information | Revelation, false attribution, falsified information |
| Anonymity of authorship | implements anonymity protections | uses forensic techniques to defeat anonymity |
| Anonymity of association | implements anonymity protections | uses forensic techniques to defeat anonymity |
| Integrity of information | assures integrity of information | defeats integrity protections, allow modification of data |
| Integrity of attribution | assures integrity of attribution | defeats integrity protections and permit falsification |
| Inviolability-safety from punishment for expression | blocks attribution, content analysis triggering retribution | defeat protections to allow attribution and retribution for speech and association |

Cybersecurity can help and hinder free expression and association. Balancing these can be essential given the growth in attacks on the inviolability for speech and association in more and more nations.

16 Barry Friedman and Maria Ponomarenko, Democratic Policing, 90 NYU L.Rev.6 (December, 2015).

17 Timothy Shilston, "Democratic Policing, Community Policing and the Fallacy of Conflation in International Police Development Missions." *International Journal of Police Science & Management* 17, no. 4, 2015.

## i. Cybersecurity that Protects Freedom of Speech

The Truman Center posits that cybersecurity can and must be configured to protect freedom of speech and personal privacy in order to enhance national cybersecurity.[18] It rejects the proposition that security can only come from ceding all personal privacy to others. A key feature for coordination of national cybersecurity among all organizations, especially those attacked, is through the sharing of information but using anonymization techniques for that information sharing.

Conversely, requiring only attributable communications will stifle freedom of speech. And become the model and justification for other countries to do so, in some cases stifling democratic movements. Implementing anonymous data and attack sharing will both enhance national cybersecurity and protect free expression and privacy rights. US courts have repeatedly struck down legislative efforts to prohibit anonymous communications, noting their importance in highlighting unpopular opinions subject to criminal sanctions.

In many ways this reflects the interplay between privacy and personal autonomy with free expression and association. With the huge embrace of ICT by nearly everyone, especially with the ubiquitous cellular telephone and its constant connection to voice, text, the Internet and any number of app programs. Whether from direct attribution of particular content or indirect inferences from traffic analysis or social network analysis, protection of speakers is essential to the free exchange of ideas.

Consider these aspects of technical systems for protecting speech and the speaker:

1. *Protects interference and modification of the information*- cyber security systems hashing or encrypting speech provide validation mechanism that tell when the content of the speech has been modified or prevent modification by blocking access to that content.
2. *Implements anonymity protections of authorship, speaker and association* – virtual private networks, encryption and obfuscation systems, including Tor implementations, help protect anonymity by stopping traceback efforts to connect speech to a particular speaker or associations between speaker and the persons receiving that speech.
3. *Assures integrity of information* – digital signatures, hashing and encryption stop or alert to attacks against the integrity of the speech and its modifications.

18 Rhodes, Matt, "Cybersecurity That Protects Freedom of Speech," Truman Center Doctrine Blog, http://trumancenter.org/doctrine-blog/cybersecurity-that-protects-freedom-of-speech/ accessed 8 June 2021.

4. *Assures integrity of attribution* – digital signatures and hashing authenticate authorship and attribution.
5. *Blocks attribution and content* analysis – VPNs and encryption may block attribution of anonymous speech and block unauthorized content analysis, especially through national firewall filters.

Similarly, cyber security can block efforts to compromise the technology of speakers as to surveil their activities, speech and associations.

Cybersecurity is required to protect free expression. By assuring the confidentiality, integrity and availability of ICT communications, and protections against inferential content and associations, the rights to speak and associate, with some level of inviolability and protection. Without any level of cyber security a speaker is open to all attacks.

But where it is only a specialized system in the hands of others, that protection is necessarily in their control and not that of the speaker. That person is, effectively, as the mercy of those others who may or may not have their interest at heart, especially if those others disagree with the speech.

## ii. Cybersecurity that Threatens Freedom of Speech

Unfortunately, the growth in sophistication of cyber-attacks requires growing sophistication in the means to combat them. Those tools, which alert to hostile cyber activity, may play a similar role in alerting to what some may feel are hostile ideas.

Consider these aspects of cyber security than may threaten speech and the speaker:

1. *Revelation, false attribution, falsified information*- cybersecurity systems that log and track activity may erroneously attribute content and authorship, leading to retribution
2. *Forensic techniques to defeat anonymity* – The growth in cyber forensics of incident detection and response, including law enforcement action, may also defeat anonymity via tracebacks, data recovery and analysis
3. *Integrity protections and modification of data* – decryption and man-in-the-middle techniques may overcome integrity protections as to permit modification of data and access to metadata for attribution
4. *Integrity protections and falsification*- decryption and man-in-the-middle techniques may overcome integrity protections as to permit falsification of data and access to/falsification of metadata for attribution

5. *Forensic discovery of authorship and association* – cybersecurity systems that log and track activity and recover and analyze data may lead to the discovery of authorship and attribution and of associations with the speaker.

Those technologies and techniques that protect may also be used to attack speech, especially in technologically-sophisticated authoritarian regimes. But those risk exist wherever technology is available, even under legal regimes that purportedly protect freedom of expression and association. These are threats that, as noted by the Truman Center, must be recognized and, if possible, remediated. Whether in law of technology.

## VI. AI and Security

Security by obscurity played a role with the protection of those expressing unpopular views as well as in personal privacy and cyber security. The rise of computing has changed that, beginning with searchable databases. And that change has grown exponentially with the rise of artificial intelligence. Pattern recognition and artificial intelligence are machine systems for finding or inferring patterns and relationships in data. The power of these systems and their deployment across multiple social, commercial and government domains impacts everyone. The risks of unintended injury from the systems is significant. Indeed, the use of facial recognition technologies in criminal justice has had to face the risks of error and injustice due to error rates in these systems, sometimes due to partial training sets that do not represent the diversity of the world.

Artificial Intelligence can detail matters private and hidden via inferences. Justice Sotomayor speculates that new, data-driven investigative technologies might change the relationship between government and the governed as it negates and shrinks the spheres of our private lives. See, e.g., *Jones v United States*.

These machine powers impact freedom of expression and association inferential analytics. AI-driven image analysis may lead to misidentification of a speaker or his associates. And subsequent injury of misattribution, mis-association and casting a person in a false light as to speech and beliefs.

In order to support "responsible development and use of AI," the Global Partnership on Artificial Intelligence of 14 countries and the European Union, aided by the OECD[19] and UNESCO, stated

19 4 Organization for Economic Cooperation and Development, "OCED to host Secretariat of new Global Partnership on Artificial Intelligence," 15 June 2020,

> Recognizing the need for cooperation at international level if we are to tap the full potential of artificial intelligence (AI) and ensure that it is of benefit to all citizens while respecting democratic values and the primacy of human beings, the founding members of the Global Partnership on Artificial Intelligence (GPAI) mean to encourage and guide responsible development of AI based on human rights, inclusion and diversity while fostering innovation and economic growth. [20] [5]

The power of AI and machine learning pushes the powers of surveillance and inference of actions and associations to unprecedented heights, at a relatively low price-point. As noted in *United States v Carpenter* and *United States v. Jones*, the consequences of exceptionally inexpensive data-driven surveillance, that in the past may have required extensive police resources, makes it possible for the surveillance of any "person of interest." Applied to matters of speech and association, there is no longer any anonymity for anyone.

## Conclusion

Cyber security protects against malicious use of cyber technologies. It protects values of a democratic society and supports the democratization ideal.[21] We must promote the democratization of cyber security for the protection of free expression and association.[22, 23, 24, 25]

---

https://www.oecd.org/going-digital/ai/OECD-to-host-Secretariat-of-new-Global-Partnership-on-Artificial-Intelligence.htm, accessed 28 June 2020.

20 Gouvernement de France, "Launch of the Global Partnership on Artificial Intelligence," 17 June 2020, https://www.gouvernement.fr/en/launch-of-the-global-partnership-on-artificial-intelligence#:~:text=The%20Global%20Partnership%20on%20AI,the%20United%20States%20of%20America, accessed 28 June 2020.

21 Ismail, Nick "Democratising technology for cyber security and the power of AI," Information Age, 24 June 2019 https://www.information-age.com/democratising-technology-cyber-security-123482856/, accessed 20 May 2021.

22 Liwer, Dror, "Now More Than Ever, We Must Democratize Cybersecurity", SmallBizDaily, 26 February 2019, https://www.smallbizdaily.com/now-more-than-ever-we-must-democratize-cybersecurity/, accessed 20 May 2021.

23 Goldfarb, Joshua, "It's Time to Democratize Security", Security Week, 9 October 2017, https://www.securityweek.com/its-time-democratize-security, accessed 20 May 2021.

24 Barlow, Caleb "It's Time for the Democratization of Cyber security Data, Security Intelligence", 21 November 2016, https://securityintelligence.com/its-time-for-the-democratization-of-cybersecurity-data/, accessed 20 May 2021.

25 Jacoby, Ari, "The future of account security must be democratized", Security, 17 December 2020.

Public safety under the democratic accountability implicit in the American Constitution are vital to the liberties and protections of citizens.[26] A subset is the concept of cyber security that is now embedded in our social relationships. Cyber security is essential to protecting rights, yet it has not been subject to either full democratic accountability nor distributed responsibility. We should consider the consequences if that does not change.

Security is security. Some have suggested the creation of a national firewall – a series of controlled access points for all Internet traffic in and out of a country – offers finite and defined transit for all electronic communications using that network. This applies regardless of transmission protocol or system. Technology now exists for the blockage of certain types or origins of communications, and, increasingly, the analysis of those communications as to protocol, content and metadata, especially social network analysis. It can serve to protect the country from malicious use of the system, such as from hostile code and malware. But the risk is its application to ideas and associations deemed harmful to the state.

This technical monitoring solution may negate the promise of the Internet to bring us together over new ideas. The Internet and computing are now so ubiquitous that they still offer great opportunities for the exchange of ideas, even under state and private attacks. A better solution is distributed technologies of cyber security for use by people. These can protect those opportunities, and those brave enough to speak on unpopular matters as to, perhaps, build a better world.

26 Barry Friedman and Maria Ponomarenko, Democratic Policing, 90 NYU L.Rev.6 (December, 2015).

**Morgan Cloud, Atlanta**

# Analog Rights, Expressive Digital Property, and a First Amendment Exclusionary Rule

## I. Introduction

At one time, Fourth Amendment rules allowed Justice Brandeis to argue "that the [Fourth] Amendment is violated by the officer's reading the paper without a physical seizure, without his even touching it."[1] He wrote at a time when constitutional rules existed to discourage government agents from seizing and reading most private papers. Were those rules enforced today, government searches of digital devices would regularly violate the Amendment.

Government agencies possess and employ tools enabling them to search our digital devices. As these searches become more common and better known, outrage gradually gives way to acceptance and people become increasingly numb to the erosion of privacy in our digital lives.[2] One result is an increasing acceptance of government searches of information stored on personal digital devices for evidence of criminal conduct.

The capabilities of government actors to search digital devices continue to evolve in tandem with increasingly sophisticated technological tools employed by criminals. In the same week in June 2021, for example, the FBI announced the recovery of millions of dollars of bitcoins paid in a ransomware attack,[3] and an international sting operation in which criminals in at least 45 countries were arrested because they had relied on encrypted devices actually supplied by the FBI.[4]

Despite the obstacles that digital security technologies like encryption continue to pose for law enforcers, digital devices remain a ready source of

---

1 Olmstead v. United States, 277 U.S. 438, 478 (1928) (Brandeis, J., dissenting).

2 *See generally*, Shoshanna Zuboff, The Age of Surveillance Capitalism: The Fight for a Human Future at the New Frontier of Power (2019).

3 Katie Benner & Nicole Perlroth, *U.S. Seizes Share of Ransom From Hackers in Colonial Pipeline Attack*, N.Y. Times (June 7, 2021), https://www.nytimes.com/2021/06/07/us/politics/pipeline-attack.html.

4 Yan Zhuang, Elian Peltier, & Alan Feuer, *The Criminals Thought the Devices Were Secure. But the Seller Was the F.B.I.*, N.Y. Times (June 9, 2021), https://www.nytimes.com/2021/06/08/world/australia/operation-trojan-horse-anom.html.

information for government searchers. "Despite claims from Apple, Google and even the Justice Department that smartphones are largely impenetrable, thousands of law enforcement agencies have tools that can infiltrate the latest phones to extract data."[5] One Israeli company, Cellebrite, "has sold data extraction tools to more than 5,000 law enforcement agencies, including hundreds of small police departments across the United States." That company's CEO states that "The solutions are there. There is no real challenge to accessing the data."[6]

Another Israeli company, NSO Group, markets spyware named Pegasus that "can collect emails, call records, social media posts, user passwords, contact lists, pictures, videos, sound recordings and browsing histories, . . . . can activate cameras or microphones to capture fresh images and recordings[,] can listen to calls and voice mails[,] can collect location logs of where a user has been and also determine where that user is now."[7]

NSO claims that it licenses Pegasus spyware "only for use in surveilling terrorists and major criminals,"[8] but an international investigation found that it had been used to hack "37 smartphones belonging to journalists, human rights activists, business executives and two women close to murdered Saudi journalist Jamal Khashoggi," and that these phones "appeared on a list of more than 50,000 numbers that are concentrated in countries known to engage in surveillance of their citizens and also known to have been clients of the Israeli firm, NSO Group."[9]

---

5 Jack Nicas & Michael S. Schmidt, *Bitcoin and Encryption: A Race Between Criminals and the F.B.I.*, N.Y. Times (June 12, 2021), https://www.nytimes.com/2021/06/12/technology/fbi-bitcoin-ransom-encryption.html.

6 *Id.* (Purchases of tools for hacking phones by U.S. police departments has reached the point where "[a]t least 2,000 law enforcement agencies in all 50 states have such tools, including 49 of the 50 largest U.S. police departments.").

7 Craig Timberg, Reed Albergotti, & Elodie Gueguen, *Despite the hype, iPhone security no match for NSO spyware*, Wash. Post (July 19, 2021), https://www.washingtonpost.com/technology/2021/07/19/apple-iphone-nso/.

8 Dana Priest, Craig Timberg, & Souad Mekhennet, *Private Israeli spyware used to hack cellphones of journalists, activists worldwide*, Wash. Post (July 18, 2021, 8:15 PM), https://www.washingtonpost.com/investigations/interactive/2021/nso-spyware-pegasus-cellphones/.

9 *Id.* The report notes that "NSO describes its customers as 60 intelligence, military and law enforcement agencies in 40 countries, although it will not confirm the identities of any of them . . . . The consortium found many of the phone numbers in at least 10 country clusters, which were subjected to deeper analysis: Azerbaijan, Bahrain, Hungary, India, Kazakhstan, Mexico, Morocco, Rwanda, Saudi Arabia and the United

It is no secret that these technological developments pose challenges for application of conventional Fourth Amendment doctrine, which is increasingly permeable to government intrusions. Not only is reading information stored on digital devices accepted more readily, but the erosion of the Fourth Amendment exclusionary rule in recent years makes it more likely that this information will be available to government prosecutors. This essay is intended as a modest experiment in reimagining how a distinct constitutional doctrine, the First Amendment right to receive ideas, might be coupled with the Fourth Amendment to provide more definite protections for some contents of digital devices by restricting their use in criminal cases.

## II. Digital "Papers" and the Intersection of First and Fourth Amendment Rights

At one time, Fourth Amendment theory imposed both substantive and procedural restrictions on government searchers' access to the contents of private papers.[10] In the late twentieth century, however, the Supreme Court eliminated the substantive protections, concluding that the procedural limits of the warrant process offered sufficient protection of constitutional rights.[11] In this chapter I examine how the intersection of First and Fourth Amendment values might provide a model for substantive protections for some communicative property. In particular, digital records stored in computers located in a person's home may serve as a model for creating heightened constitutional protection even when a warrant authorizes government agents to seize and search the device.

The Supreme Court's well-known decision in *Riley v. California*, emphasized the unique characteristics of contemporary digital devices – in that case, cell phones – and the information stored in them.[12] The technical Fourth Amendment issue in *Riley* was whether a particular Fourth Amendment doctrine, the search incident to arrest exception to the warrant requirement, justified warrantless searches of cell phones. In an opinion written by Chief Justice Roberts, the Court held that it did not.[13] The justifications for warrantless searches of physical

---

Arab Emirates. Citizen Lab also has found evidence that all 10 have been clients of NSO." *Id.*

10 *See e.g.*, Boyd v. United States, 116 U.S. 616 (1886); Weeks v. United States, 232 U.S. 383 (1914); Silverthorne Lumber Co. v. United States, 251 U.S. 385 (1920).

11 Warden v. Hayden, 387 U.S. 294 (1967); United States v. Doe, 465 U.S. 605 (1984).

12 Riley v. California, 573 U.S. 373, 378–82.

13 *Id.* at 386.

objects did not have "much force with respect to digital content on cell phones"[14] because of the nature of modern digital technologies, which can store vast quantities of data, even on a cell phone with comparatively limited storage. A device with merely 16 gigabytes of storage capacity "translates to millions of pages of text, thousands of pictures, or hundreds of videos."[15] The capacity to store quantities of data of diverse content makes even these mini computers unique for Fourth Amendment purposes.

> One of the most notable distinguishing features of modern cell phones is their immense storage capacity. Before cell phones, a search of a person was limited by physical realities and tended as a general matter to constitute only a narrow intrusion on privacy. Most people cannot lug around every piece of mail they have received for the past several months, every picture they have taken, or every book or article they have read.[16]

The quantity and variety of information that can be stored digitally make it possible for searchers to reconstruct "an individual's private life" from the photographs, stored messages, lists of contacts, records of phone calls, music and video libraries, the applications installed in the phone, as well as data stored in the "cloud"[17] accessible from the device.[18] Inevitably, the nature of this digital record is *qualitatively* different from other forms of physical record keeping.

> An Internet search and browsing history, for example, can be found on an Internet-enabled phone and could reveal an individual's private interests or concerns—perhaps a search for certain symptoms of disease, coupled with frequent visits to WebMD. Data on a cell phone can also reveal where a person has been. Historic location information is a standard feature on many smart phones and can reconstruct someone's specific movements down to the minute, not only around town but also within a particular building.[19]

Despite this extensive explanation of the differences between records stored on digital devices and those stored in traditional media like physical paper records, the Supreme Court ultimately offered the same limited procedural protection for both. A judicial search warrant justified the search of the digital storehouse as much as it would a single paper file.

---

14 *Id.*; *see also id.* at 393 ("Modern cell phones, as a category, implicate privacy concerns far beyond those implicated by the search of a cigarette pack, a wallet, or a purse.").

15 *Id.* at 394.

16 *Id.* at 393 (citations omitted).

17 *Id.* at 397.

18 *Id.* at 393.

19 *Id.* at 395–96.

I propose that theories offering protection for analog forms of information – theories arising at the intersection of the First and Fourth Amendments – might be applied to shield some digital records from government intrusions, even those authorized by valid search warrants.

The two amendments are organically linked. The Fourth Amendment guarantees the people's right to be secure in our "persons, houses, papers, and effects."[20] The inclusion of "papers" in the list of protected property rights would be redundant – papers are effects – if the purpose was only to protect physical objects. Papers are singled out because of their unique nature as communicative property, both the expression and the repository of the people's ideas, thoughts, values, and emotions. This is consistent with the Amendment's historic roots in British cases involving government searches and seizure of private papers, and in the Lockean property theories embedded in the Amendment's text.[21]

In recent decades, scholars have recognized the historic connections between this element of the Fourth Amendment and the First Amendment's guarantees of freedom of thought and expression.[22] Taken together, the two Amendments can be seen as erecting substantive as well as procedural barriers to government efforts to search for, seize, and use the contents of private papers against their owners and creators.[23]

---

20 U.S. CONST. amend. IV.

21 *See* Morgan Cloud, *Privacy is Property*, 55 AM. CRIM. L. REV. 37, 42–55 (2018).

22 *See, e.g.*, William J. Stuntz, *The Substantive Origins of Criminal Procedure*, 105 YALE L.J. 393, 395 (1995) (there is a "strong tradition of using Fourth and Fifth Amendment law as a shield against government information-gathering—a tradition that has more to do with protecting free speech than with regulating the police."); Paul Belonick, *Transparency is the New Privacy: Blockchain's Challenge for the Fourth Amendment*, 23 STAN. TECH. L. REV. 114, 162–63 (2020) ("Any attempt to grapple with the Fourth Amendment's meaning must take into account its original close relationship with the First Amendment."); Michael W. Price, *Rethinking Privacy: Fourth Amendment "Papers" and the Third-Party Doctrine*, 8 J. NAT'L SECURITY L. & POL'Y 247, 257–58 (2016) ("[A] paramount purpose of the Fourth Amendment was to serve as a guardian of individual liberty and free expression . . . . intended to function as a barrier to government overreach and as a catalyst for other constitutional rights, notably freedom of speech and freedom of association.").

23 Neil Richards, *The Third Party Doctrine and the Future of the Cloud*, 94 WASH. U. L. REV. 1441, 1450–51 (2017) (the Fourth Amendment's specific reference to "papers" serves as a "substantive limit on government power primarily in the context of communications and dissent.").

This is consistent with Fourth Amendment doctrines in place from the late nineteenth century to the second half of the twenty-first century. As described by Justice Douglas, dissenting in the case that abandoned substantive Fourth Amendment rights, this traditional view held that the Fourth Amendment erected two types of protection against government intrusions; one was substantive, the other procedural:

(1) One creates a zone of privacy that may not be invaded by the police through raids, by the legislators through laws, or by magistrates through the issuance of warrants.
(2) A second creates a zone of privacy that may be invaded either by the police in hot pursuit or by a search incident to arrest or by a warrant issued by a magistrate on a showing of probable cause.[24]

The substantive protections were first articulated in a case holding that the Fourth and Fifth Amendments protected private commercial papers from government searches and seizures. Unless private papers fell into a limited set of categories – they were stolen property, contraband, instrumentalities used to commit crimes, or certain "required records" – the Fourth and Fifth Amendments read individually and together prohibited the government from seizing private papers and using them as evidence in criminal or quasi-criminal cases,[25] even if the search and seizure was authorized by a valid warrant.[26]

The Constitution offered this protection precisely because papers are communicative property – they express ideas, and "any forcible and compulsory extortion of a man's own testimony, or of his private papers . . . to convict him of crime, or to forfeit his goods, is within the condemnation of that judgment. "In this regard, the Fourth and Fifth Amendments run almost into each other."[27] *Boyd's* holding was understood to create a ban on paper searches, a constraint that proved unsustainable in twentieth century investigations.[28] But which papers can be searched remains a Fourth Amendment issue, particularly in an age of digital documents.

By any rational definition, digital documents are the contemporary equivalent of physical papers.[29] Like papers, their contents may deserve special protection

24 Warden v. Hayden, 387 U.S. 294, 313 (1967) (Douglas, J., dissenting).

25 Boyd v. United States, 116 U.S. 616, 624 (1886).

26 Weeks v. United States, 232 U.S. 383, 390–94 (1914).

27 *Boyd*, 116 U.S. at 630.

28 United States v. Doe, 465 U.S. at 618 (O'Connor, J., concurring).

29 *See, e.g.*, Carpenter v. United States, 1338 S. Ct. 2206, 2230 (2018) (Kennedy, J., dissenting) (The Fourth Amendment third party doctrine "may not apply when the

under both the First and Fourth Amendments. "[T]he Fourth Amendment must be applied with 'scrupulous exactitude' when significant First Amendment rights are implicated by a search. It follows, scholars have argued, that in the digital realm the Fourth Amendment protects more than privacy; it protects a right to digital expression and association in all manner of digital communications among persons."[30]

In *Riley*, the Supreme Court affirmed that the Fourth Amendment protects the expressive property stored in cell phones, even when the devices are carried in public. "With all they contain and all they may reveal, they hold for many Americans 'the privacies of life.' The fact that technology now allows an individual to carry such information in his hand does not make the information any less worthy of the protection for which the Founders fought."[31]

When these electronic records are stored on devices located in the home, the same Fourth Amendment concerns not only exist but are amplified, because the home is the most protected place under the Fourth Amendment.[32] The First Amendment imposes complementary substantive limits on the government's authority to search for, seize, and use some information stored on electronic devices, particularly when kept in the home. These First Amendment substantive

---

Government obtains the modern-day equivalents of an individual's own 'papers' or 'effects,' even when those papers or effects are held by a third party", and listing "letters held by mail carrier" and "e-mails held by Internet service provider" as examples.); *id.* at 2269 (Gorsuch, J., dissenting) ("Just because you entrust your data—in some cases, your modern-day papers and effects—to a third party may not mean you lose any Fourth Amendment interest in its contents . . . . [F]ew doubt that e-mail should be treated much like the traditional mail it has largely supplanted — as a bailment in which the owner retains a vital and protected legal interest.").

30 Paul Belonick, *Transparency is the New Privacy: Blockchain's Challenge for the Fourth Amendment*, 23 Stan. Tech. L. Rev. 114, 162–63 (2020) (citing Supreme Court decisions), Stanford v. Texas, 379 U.S. 476, 485 (1965); Zurcher v. Stanford Daily, 436 U.S. 547, 564 (1978); New York v. P. J. Video, Inc., 475 U.S. 868, 873 (1986), and Fourth Amendment scholarship: Price, *supra* note 23, at 283–86 ("[T]he content of communications, whether spoken, typed, or beamed over the Internet, are the kind of expressive and associational materials that the Framers intended to shield from arbitrary search and seizure through the Fourth Amendment."), and Richards, *supra* note 24, at 1487–88 ("In translating the Fourth Amendment to the cloud, . . . Fourth Amendment protection should be strongest when dealing with social and technological activities that are intertwined with a First Amendment value.").

31 Riley v. California, 573 U.S. at 403 (quoting Boyd v. United States, 116 U.S. 616, 625, 630 (1886)).

32 *See* Weeks, 232 U.S. at 390.

rights exist and can provide constraints on government searchers independent of contemporary Fourth Amendment procedural requirements.

## III. The First Amendment and Possession of Obscene Materials in the Home

No case better illustrates the intersection of the First and Fourth Amendments in searches of expressive property stored in a home than does *Stanley v. Georgia*.[33] Federal and state agents, executing a valid warrant, searched Stanley's home for evidence of bookmaking crimes. They discovered no evidence of gambling activities, but during the search they found three pornographic movies. Stanley was indicted and convicted for "knowingly hav[ing] possession of . . . obscene matter" prohibited by Georgia state law.[34] Stanley filed a pretrial motion to suppress the films, arguing that "they had been seized in violation of the Fourth and Fourteenth Amendments."[35] The state lower courts rejected this claim, and the Georgia Supreme Court "specifically determined that the films had been lawfully seized."[36]

All nine Supreme Court justices voted to reverse Stanley's conviction. The majority held that the seizure of the films and Stanley's prosecution and conviction for possession of obscene materials violated his First Amendment rights. Three justices concurred in the result but on the grounds that the seizure of the films violated the Fourth Amendment. The justices' analyses under both Amendments have implications for contemporary searches of the contents of digital devices.

### A. The First Amendment Right to Receive Information and Ideas

Two First Amendment issues in the case were not disputed. Stanley did *not* dispute that the films were obscene, and the Supreme Court majority assumed "they are obscene under any of the tests advanced by members of this Court."[37] The majority also concluded that under its prior decisions, "obscenity is not protected by the First Amendment."[38]

---

33 394 U.S. 557 (1969).

34 *Id.* at 558.

35 *Id.* at 569 (Stewart, J., concurring).

36 *Id.*

37 *Id.* at 570 n.2.

38 *Id.* at 560.

But the majority focused on another question not addressed in earlier decisions: Could the state criminalize "mere private possession of obscene material"?[39] Its prior decisions involved "the power of the State and Federal Governments to prohibit or regulate certain public actions" like the sale or distribution of obscene materials.[40] Georgia argued that it was empowered to prosecute people for the possession of materials the state deemed harmful to its citizens' welfare. "If the State can protect the body of a citizen, may it not, argues Georgia, protect his mind?"[41]

Not from receiving ideas and not from keeping them in the privacy of the home:

> It is now well established that the Constitution protects the right to receive information and ideas. 'This freedom (of speech and press) . . . necessarily protects the right to receive . . . .' This right to receive information and ideas, regardless of their social worth, is fundamental to our free society. Moreover, in the context of this case—a prosecution for mere possession of printed or filmed matter in the privacy of a person's own home — that right takes on an added dimension. For also fundamental is the right to be free, except in very limited circumstances, from unwanted governmental intrusions into one's privacy.[42]

A fundamental First Amendment right – to receive ideas[43] regardless of their social value – was amplified when those ideas were preserved as expressive property and kept and used in the privacy of one's home. The Court's justification for this conclusion was framed expansively, and in terms with provocative implications for government authority to search for and use the contents of digital devices, including those kept in the home.

39 *Id.* at 559.
40 *Id.* at 561– 62.
41 *Id.* at 560.
42 *Id.* at 564 (citations omitted).
43 The Supreme Court first recognized the right to receive information in *Martin v. Struthers*. 319 U.S. 141 (1943). In *Board of Education v. Pico*, the Court described this right "as an inherent corollary of the rights of free speech and press that are explicitly guaranteed by the Constitution, in two senses. First, the right to receive ideas follows ineluctably from the sender's First Amendment right to send them . . . . More importantly, the right to receive ideas is a necessary predicate to the recipient's meaningful exercise of his own rights of speech, press, and political freedom." 457 U.S. 853, 867 (1982). In *Griswold v. Connecticut*, the Court emphasized that without the right to receive, the First Amendment itself would be less secure. 381 U.S. 479, 482–83.

> These are the rights that appellant is asserting in the case before us. He is asserting the right to read or observe what he pleases — the right to satisfy his intellectual and emotional needs in the privacy of his own home. He is asserting the right to be free from state inquiry into the contents of his library. Georgia contends that appellant does not have these rights, that there are certain types of materials that the individual may not read or even possess. Georgia justifies this assertion by arguing that the films in the present case are obscene. But we think that mere categorization of these films as 'obscene' is insufficient justification for such a drastic invasion of personal liberties guaranteed by the First and Fourteenth Amendments. Whatever may be the justifications for other statutes regulating obscenity, we do not think they reach into the privacy of one's own home. *If the First Amendment means anything, it means that a State has no business telling a man, sitting alone in his own house, what books he may read or what films he may watch. Our whole constitutional heritage rebels at the thought of giving government the power to control men's minds.*[44]

These rights are not unlimited,[45] but they are fundamental, embedded in the First Amendment, and imposed on the states by the Fourteenth. The breadth of this holding suggests the possibility of substantive limits not only on searches for digitally stored information, but also its use against a criminal suspect.

## B. The Right to Receive and Possess Seditious Materials

Consider, for example, the ongoing investigations of the violent insurrection against the United States government at the Capitol on January 6, 2021. The investigation of crimes committed by participants has relied heavily on digital sources of information, ranging from social media postings, to websites maintained by political groups and individuals, to electronic devices seized from individual suspects.

Law enforcers' access to and use of the information posted on social media or political websites would generally be unaffected by the holding in *Stanley*. For example, a video posted on social media memorializing people's actions, including their spoken words, would remain readily available to searches by private citizens

---

44 *Id.* at 565 (citations omitted) (emphasis not in original).

45 In *Osborne v. Ohio*, for example, the Supreme Court distinguished *Stanley's* "narrow holding", and held that the "the interests underlying child pornography prohibitions far exceed the interests justifying the Georgia law at issue in *Stanley*." 495 U.S. 103, 108 (1990). The state was justified in criminalizing possession of child pornography to decrease demand as part of its efforts to 'decrease the production of child pornography' and destroy the market for it. *Id.* Intrusive investigative methods were also justified because child pornography has 'been driven underground' making it much harder for law enforcement to intervene in the market. *Id.* at 110.

and government agents alike, and for use as evidence in criminal prosecutions.[46] Neither visual recordings of criminal conduct nor verbal admissions of criminal acts distributed publicly are insulated by First Amendment principles from government use. Similarly, information posted on publicly accessible sites is not protected by Fourth Amendment privacy principles.

But some information stored on electronic devices might warrant protection under the holding of *Stanley*. Imagine that government searchers seize a computer from a suspect's home that contains files that advocate radical political doctrines – violent revolution, overthrowing the United States government, and hanging certain political leaders, including the Vice President. The owner and user of the device in this hypothetical has been charged in a federal indictment with committing multiple crimes during the insurrection at the United States Capitol on January 6, 2021, and prosecutors want to use these files as evidence against the suspect, perhaps to confirm his culpable mental state during the riot.

*Stanley* suggests the following constitutional rights argument against allowing the government to use the files in evidence against the suspect. The government cannot penalize mere possession of even illegal expressive property, like obscene materials, without violating the First Amendment. The same principle should apply to privately possessed materials advocating seditious violence. Under *Brandenburg v. Ohio*, government cannot criminalize mere possession of materials containing even violent seditious ideas because "the constitutional guarantees of free speech and free press do not permit a State to forbid or proscribe advocacy of the use of force or of law violation *except where such advocacy is directed to inciting or producing imminent lawless action and is likely to incite or produce such action*."[47]

The constitutional privilege could sweep more broadly, preventing not only criminalization of possession of these materials but also prohibiting the use of privately maintained seditious materials in the government's case in chief against the January 6th insurrection suspect. Setting aside admissibility issues

---

46 *See, e.g.*, Russell L. Weaver, *Brandenburg and Incitement in a Digital Era*, 80 Miss. L. J. 1263, 1288 (2011) ("At the same time, the Internet allows for enhanced law enforcement surveillance of communications, and therefore may give the government greater capacity to observe and intervene."); *see also*, Abigail M. Pierce, *#Tweeting for Terrorism: First Amendment Implications in Using Proterrorist Tweets to Convict Under the Material Support Statute*, 24 Wm. & Mary Bill Rts. J. 251 (2015); Morgan Stacey, *Americans, ISIS, and Social Media: How the Material Support Statute Can Help Combat Against Their Collision*, 30 Notre Dame J.L. Ethics & Pub. Pol'y 201 (2016).

47 395 U.S. 444, 447 (1969) (emphasis not in original).

concerning the files' relevance and probative value in the government's case,[48] the materials should be protected by the First Amendment, and thus privileged under Rule 501 of the Federal Rules of Evidence.[49] They should not be admissible in evidence in the criminal case against their owner.

The practical impact of such a First Amendment rule of exclusion on legitimate prosecutions should be limited. Before bringing the case, it is likely the government will have a wealth of evidence in hand: video recordings of the event from a variety of sources showing the suspect at the insurrection; statements by the suspect recorded and posted on social media and other internet sites confirming his involvement in the insurrection; eyewitness accounts; and the like. But the practical implications of a substantive First Amendment prohibition on using files stored on the suspect's digital devices is relevant to the results of searches and seizures of such machines. Again, *Stanley* offers a starting point for discussion.

Justice Stewart, joined by Justices Brennan and White, concurred in the result in *Stanley*, but objected to the majority's reliance on the First Amendment rather than the Fourth. After noting that Stanley's original motion to suppress the obscene films was based on the Fourth and Fourteenth Amendments, Stewart complained that "the Court today disregards this preliminary issue in its hurry to move on to newer constitutional frontiers. I cannot so readily overlook the serious inroads upon Fourth Amendment guarantees countenanced in this case by the Georgia courts."[50]

Justice Stewart acknowledged that federal and state agents originally acted lawfully, entering Stanley's home under a valid warrant describing with particularity the place to be searched – the home – and things to be seized – "equipment, records, and other material used in or derived from an illegal wagering business."[51] But the seizure and search of the canisters containing the obscene films amounted to "the general searches and unrestrained seizures that had been a hated hallmark of colonial rule under the notorious writs of assistance of the British Crown. This most basic of Fourth Amendment guarantees was frustrated in the present case."[52]

---

48 *See* Fed. R. Evid. 401–403.

49 Rule 501 provides: "The common law — as interpreted by United States courts in the light of reason and experience — governs a claim of privilege unless any of the following provides otherwise: the United States Constitution . . . ." *Id.* 501.

50 394 U.S. at 569 (Stewart, J., concurring).

51 *Id.* at 570.

52 *Id.* at 569–570.

The agents acted legally "when they proceeded to the appellant's upstairs bedroom and pulled open the drawers of his desk. But when they found in one of those drawers not gambling material but moving picture films, the warrant gave them no authority to seize the films."[53] Their action violated a fundamental constitutional principle: "The requirement that warrants shall particularly describe the things to be seized makes general searches under them impossible and prevents the seizure of one thing under a warrant describing another. As to what is to be taken, nothing is left to the discretion of the officer executing the warrant."[54]

## IV. The Practical Limits of Fourth Amendment Particularity in Digital Searches

In the abstract, this principle is readily applicable to searches of digital devices and the scope of warrants authorizing those searches, but the nature of these searches in practice is messier. The rules governing computer searches – constitutional, statutory, and procedural – and the practical realities of these searches, are far too voluminous to address here.[55] The discussion instead focuses upon practical issues related to the use of materials arguably privileged under *Stanley* that are uncovered during searches of digital devices.

Warrants authorizing computer searches must, of course, satisfy the Fourth Amendment's probable cause and particularity requirements. Because of the vast quantities and diverse types of data that can be stored on a computer, the warrant should be specific in identifying "which type of files are sought."[56]

But the practical realities of locating data stored on computers may force searchers to inspect information that turns out to be outside the scope of the warrant. "Even if the agents know specific information about the files they seek, the data may be mislabeled, encrypted, stored in hidden directories, or embedded

53 *Id.* at 571.

54 *Id.* (citation omitted).

55 *See, e.g.*, Off. of Legal Educ., Exec. Off. for U.S. Att'ys, Dep't of Just., Searching and Seizing Computers and Obtaining Electronic Evidence in Criminal Investigations (2009), https://www.justice.gov/sites/default/files/criminal-ccips/legacy/2015/01/14/ssmanual2009.pdf. This departmental manual for Justice Department attorneys is 287 pages long, and does not include discussion of the many computer searches and seizures conducted since 2009.

56 United States v. Carey, 172 F.3d 1268, 1275 (10th Cir. 1999); *see also*, United States v. Otero, 563 F.3d 1127, 1132 (10th Cir. 2009); United States v. Riccardi, 405 F.3d 852, 862 (10th Cir. 2005); United States v. Adjani, 452 F.3d 1140, 1148 (9th Cir. 2006).

in 'slack space' that a simple file listing will ignore . . . . Moreover, evidence of a crime will not always take the form of a file. It may be in a log, operating system artifact, or other piece of recorded data that can be difficult to locate and retrieve without the appropriate tools and time."[57]

As a result, computer searchers often must explore extensively through the machines' contents to identify the information covered by the warrant. Even if Justice Brandeis was correct that the officers violate the Fourth Amendment merely by reading constitutionally protected papers,[58] such intrusions are inevitable in contemporary searches of computers and other digital storage devices. In searches related to the January 6, 2021 insurrection at the Capitol, it is all but certain that some devices will contain information advocating violent rebellion against the government. This is the point at which an exclusionary rule for information privileged by the First Amendment right to receive and keep information might have impact; not by preventing searchers from inspecting the information, but by prohibiting the government from using privileged contents to prosecute the individual.

The "plain view" doctrine illustrates why the Fourth Amendment alone may offer inadequate protections, and why a First Amendment exclusionary rule is useful in this circumstance. The plain view exception to the warrant requirement allows an agent to seize evidence in a place the agent is legally entitled to be, if the incriminating nature of the evidence is "immediately apparent" – that is, the officers have probable cause to believe the evidence is related to criminality.[59]

"The most common use of the plain view doctrine in the computer context occurs when agents examine a computer pursuant to a search warrant and discover evidence of a separate crime that falls outside the scope of the search warrant."[60] This is analogous to the scenario in our hypothetical – government searchers executing a warrant access files containing seditious materials not satisfying the *Brandenburg* standard. Under Fourth Amendment doctrine, arguably these files were discovered in "plain view," and could be seized and used as evidence against the riot suspect, perhaps to establish a culpable state of mind.

---

57 *Searching and Seizing Computers and Obtaining Electronic Evidence in Criminal Investigations*, *supra* note 55, at 76; *see also*, United States v. Hill, 459 F.3d 966, 974 (9th Cir. 2006) ("officers would have to examine every one of what may be thousands of files on a disk").

58 Olmstead v. United States, 277 U.S. 438, 478 (1928) (Brandeis, J., dissenting).

59 *See, e.g.*, Horton v. California, 496 U.S. 128, 136 (1990).

60 *Searching and Seizing Computers and Obtaining Electronic Evidence in Criminal Investigations*, *supra* note 55, at 34.

But under the First Amendment principles announced in *Stanley*, these files would not be evidence of criminality. They would be constitutionally protected ideas the suspect was entitled to receive and store. That is the import of *Stanley's* conclusion that "[i]f the First Amendment means anything, it means that a State has no business telling a man, sitting alone in his own house, what books he may read or what films he may watch."[61] Here, the First Amendment principle produces a bright line rule that resolves any ambiguity left by Fourth Amendment rules like the plain view doctrine. Unless that information satisfies the *Brandenburg* standard and is "directed to inciting or producing imminent lawless action and is likely to incite or produce such action,"[62] it is privileged, and cannot be used in evidence against the suspect.[63]

## V. Conclusion

Although the fact that the search in *Stanley* occurred in his home was emphasized by the justices in their opinions, the practical realities of twenty-first century digital life suggest that the First Amendment right and exclusionary rule I propose here should apply equally to devices seized outside the home. First and Fourth Amendment constitutional values protecting data stored on a digital device exist whether the device is located in the home or in public. Being located within a home may enhance the Fourth Amendment privacy protections, but they are not restricted to that location.[64] The quantity and quality of data stored does not vary with the location of the device, it is inherent in the nature of the mobile computer – whether a phone, a tablet, or a laptop. Ultimately, the First Amendment rights to receive and possess information should be the decisive factor.

---

61 Stanley, 394 U.S. at 565.

62 Brandenburg, 395 U.S. at 447.

63 *Searching and Seizing Computers and Obtaining Electronic Evidence in Criminal Investigations, supra* note 55, at 98–100 (discussing defendants' motions for return of computers or documents).

64 Katz v. United States, 389 U.S. 347, 351 (1967) ("What a person . . . . seeks to preserve as private, even in an area accessible to the public, may be constitutionally protected.").

Iris Nguyên Duy, Kristiansand

# Contemporary Threats to Freedom of Expression in Norway in the Context of the 2021 Parliamentary Election and the COVID-19 Pandemic

As "[t]he concept of free and uninhibited speech permeates all truly democratic societies and institutions",[1] the preservation of democracy is entwined with that of free speech.[2] It is especially important to protect the freedom of expression during election periods, because, for most people, voting by choosing a candidate or a political party and casting a ballot is the primary means of taking part in politics.[3] This specific act may even be seen as the *most potent form of free speech* since it is the one form of political expression that directly contributes to changing the political landscape in a region or a country (or even the world) for the years to come.[4]

Alas, because of the COVID-19 pandemic and the increased use of new technologies to influence the outcome of national elections, one of the recent main

1 «It is difficult to imagine a guaranteed right more important to a democratic society than freedom of expression. Indeed, a democracy cannot exist without that freedom to express new ideas and to put forward opinions about the functioning of public institutions. The concept of free and uninhibited speech permeates all truly democratic societies and institutions. The vital importance of the concept cannot be over-emphasized.» – Justice Cory of Canada's Supreme Court, Edmonton Journal v. Alberta (Attorney General), 1989 CanLII 20 (SCC), [1989] 2 SCR 1326, https://www.canlii.org/en/ca/scc/doc/1989/1989canlii20/1989canlii20.html

2 "Freedom of political expression is a necessary component of democracy." – Ashutosh Bhagwat & James Weinstein, Freedom of Expression and Democracy, in Adrienne Stone, Frederick Schauer (eds.), *The Oxford Handbook on Freedom of Speech*, Oxford University Press, 2021, Chapter 5, p. 83. Political expression is a necessary component of popular sovereignty and of formal political equality, see ibidem, respectively p. 84 f. and p. 88 f.

3 All citizens participate *directly* in government and public affairs through the casting of votes or by being elected through free and fair elections held periodically. Political participation can also take the form of referendums or plebiscites.

4 In this article, it is assumed that voting is speech: casting a ballot or even signing a petition are expressive activities, both because they are acts and have a political content.

challenges for the States worldwide has been the organisation of free and fair elections – and securing the voters' freedom of expression. According to the International Institute for Democracy and Electoral Assistance (International IDEA), at least 78 countries and territories across the globe decided to postpone national or subnational elections, between February 2020 and March 2021, due to COVID-19, whereas at least 113 decided to hold elections as originally planned.[5] Many governments have limited traditional campaigning as part of broader COVID-19 measures (typically, by banning or reducing the number of people who can attend campaign events such as party rallies or conventions, and public gatherings); Remote campaigning has increased in many countries as a result of the restrictions, especially on social media.[6] The COVID-19 pandemic has also had the following double effect on freedom of expression in the context of elections: on the one hand, the accessibility and increased use of the internet and social media during the pandemic has improved the voters' access to information and facilitated their participation to the public debate; on the other hand, it has equally improved the possibilities to reach, influence and even manipulate the voters via the same channels; not to mention the pandemic's economic, sociologic and psychological effects on the future voters.

In Norway, parliamentary elections are held every four years to renew all the 169 seats in the Norwegian legislature, the Storting.[7] As the next parliamentary election is scheduled on 13 September this year, the purpose of the article is to uncover or draw the attention to selected contemporary threats to freedom of expression specific to Norway, in this political context and that of the COVID-19 pandemic.

This paper is based on a broad understanding of freedom of political expression, that not only includes freedom of speech (whether oral or written),[8] the right to seek, receive and impart information and ideas of all kinds, by any means,

5 International Institute for Democracy and Electoral Assistance (IDEA), Global overview of COVID-19: Impact on elections, IDEA, 29 March 2021, https://www.idea.int/news-media/multimedia-reports/global-overview-covid-19-impact-elections (last accessed: 08.04.21).

6 Erik Asplund, Elections and COVID-19: How election campaigns took place in 2020, 2 February 2021, https://www.idea.int/news-media/news/elections-and-covid-19-how-election-campaigns-took-place-2020 (last accessed: 08.04.21).

7 The same day, there is also an election to choose 39 representatives to the Sami Parliament (Sametinget), which is held at the same time as the parliamentary elections.

8 "the right to contribute to public opinion by freely expressing their views on matters of public concern" – Ashutosh et al., Freedom of Expression and Democracy, *op. cit.*, p. 89.

the right to form an opinion and express it freely, but also, in the specific context of national elections, the right to participate to the election, either as a voter or as a candidate, and to cast a ballot (and eventually to contest electoral results).[9]

More “traditional” and/or long-standing global issues such as the significant power wielded by tech giants, their role as a media or not (e.g. their legal responsibility as portals to information and/or as news editors), the global “rise of the bots” and the control of the news feed on social media by algorithms (creating echo chambers, filter bubbles etc.), threats to journalists, the general difficulty of distinguishing between truth and falsity in the context of robust election debates,[10] Facebook’s “ban on praise, support and representation of white nationalism and white separatism on Facebook and Instagram”[11] and the so-called “Facebook’s Supreme Court”, the Oversight Board,[12] as well as the need (or not) to regulate electoral speech,[13] fall outside the scope of this article. The

---

9 But I will not deal with freedom of the press, the right to petition the government and freedom of political association in this paper. In the rest of the paper, «freedom of speech» will be sometimes used interchangeably with «freedom of expression», although the relationship between these two concepts is more complex. See Frederick Shauer, What is Speech? The Question of Coverage, in Adrienne Stone, Frederick Schauer (eds.), *The Oxford Handbook on Freedom of Speech*, Oxford University Press, 2021, Chapter 9.

10 Although “political speeches concern decisions about possible courses of action which are contentious and contested and about which people might reasonably disagree”, the idea of free speech does not justify the protection and spreading of lies. – Political Studies Association, Writing a Political Speech, https://www.psa.ac.uk/what-makes-speech-political

11 Facebook, Standing against Hate, 27.03.19, https://about.fb.com/news/2019/03/standing-against-hate/ The Norwegian Commission for Freedom of Expression is currently investigating measures against “hateful, discriminatory and in other ways offensive statements” and assessing the question of the “role and legal responsibility” of social media. See the Commission’s mandate: https://www.regjeringen.no/contentassets/92f1eb72e0534500a31ac07ddc3f8e3d/ytringsfrihetskommisjon---mandat.pdf

12 Official website: https://oversightboard.com/ See David Kayes’ preliminary reaction to the creation of the Board here https://www.ohchr.org/_layouts/15/WopiFrame.aspx?sourcedoc=/Documents/Issues/Opinion/Legislation/OL_OTH_01_05_19.pdf&action=default&DefaultItemOpen=1, and a first evaluation here: https://www.ohchr.org/_layouts/15/WopiFrame.aspx?sourcedoc=/Documents/Issues/Opinion/ResearchPaper2020.pdf&action=default&DefaultItemOpen=1

13 Jacob Rowbottom, Lies, Manipulation and Elections — Controlling False Campaign Statements, *Oxford Journal of Legal Studies*, Vol. 32, No. 3 (2012), pp. 507–535, doi:10.1093/ojls/gqs016.

same applies to the issues of misinformation and disinformation on/about the COVID-19 pandemic ("infodemic") and their impact on freedom of speech either, although they equally affect decision-making processes and opinions (in a mainly negative way).[14]

The legal standards and measures (regulatory and others) taken by the Norwegian authorities to prevent or tackle any infringement of freedom of expression in this specific context will be presented, and the weaknesses or shortcomings of the existing legal framework will be highlighted where relevant.

Freedom of expression implies the right to seek and receive information and ideas to form an informed opinion and to participate to the matters of State. The first section of the paper focuses on the digital challenges (both national and foreign) to this aspect of freedom of expression in the specific context of parliamentary elections.

Freedom of expression also entails the right to impart (convey, communicate) information and ideas, to express oneself and participate in the public debate. The second section of the paper focuses on digital threats and hate speech, and on their chilling effect on both the victims of hate speech and those who wish to express themselves unhindered.

As freedom of expression also covers the right to participate in the electoral debate and, ultimately, to cast a ballot, the third section of the paper focuses on some specific challenges created by the Covid 19-pandemic in the specific context of the 2021 parliamentary election – and on how they are dealt with by the Norwegian authorities.

## 1. The Main Digital Threats to Freedom of Expression and Fair Elections

There are at least two issues linked to digital access to information: a first issue is the content and quality of the information disseminated and received; another

14 See, for example, OECD, Combatting COVID-19 disinformation on online platforms, 03.07.21, https://read.oecd-ilibrary.org/view/?ref=135_135214-mpe7q0bj4d&title=Combatting-COVID-19-disinformation-on-online-platforms&_ga=2.47720239.980231088.1622031841-852156249.1622031841; Department of Evidence and Intelligence for Action in Health, Understanding the Infodemic and Misinformation in the fight against COVID-19, Factsheet nr. 5, Pan American Health Organization, 2020, https://iris.paho.org/bitstream/handle/10665.2/52052/Factsheet-infodemic_eng.pdf?sequence=14 (last accessed: 26.05.21).

(partially overlapping) issue is the kind of information the electorate receives and in what quantity.

With the development of new technologies and of the digital media, people have more content to choose from, and from an almost unlimited number of sources. But a direct challenge to this social and technological evolution is that it is more difficult to distinguish between what is true or false, to understand where the information comes from and whether it is reliable, in a complex media landscape and an overflow of information.

Information of quality (i.e., reliable, trustworthy information) is vital to form an opinion and to make enlightened choices. During an election year, it is even more important for an *informed voting*. False statements during an election campaign can undermine the integrity of an election.[15] It is mainly this aspect of the freedom of expression that either suffers from, or is targeted by, diverse digital threats, mostly *during* the electoral campaign.[16] The result is a manipulated opinion with, *inter alia*, potentially damaging consequences for the casting of ballots (the outcome of the election can be distorted if enough voters are manipulated or misled into changing their preferences).

As governmental transparency does not constitute a major contemporary threat to freedom of expression in Norway,[17] this section will mainly focus on

---

15 See Jacob Rowbottom, Lies, Manipulation and Elections, *op. cit.*, pp. 509–519.

16 But digital hacking and manipulation can of course happen during the vote and after the vote, during the counting of ballots or in the form of speech spread wide, contesting the result of the elections and sawing doubt in the electors' minds.

17 Public bodies have an obligation to disclose information and every member of the public has a corresponding right to receive information. As a rule, public bodies publish and disseminate widely documents of significant public interest. Openness in public administration is regulated in the Public Government Act (*forvaltningsloven*), the Freedom of Information Act (*offentlighetsloven*) and the Environmental Information Act (*miljøinformasjonsloven*). To make it easier to gain access to public documents, thereby contributing to greater openness in public administration, the Agency for Public Management and eGovernment (Difi) has developed the new access solution e-access. E-access makes it possible to follow political decision-making processes across organisational boundaries. And, like the other Nordic countries, Norway has a long tradition of civil society actively participating in decision-making processes. See the Norwegian governmental website dedicated to openness at https://open.regjeringa.no/ Norway joined the Open Government Partnership (OGP) in 2011 and performs "among the top countries in the world on measurements of good governance, transparency, and anti-corruption". See https://www.transparency.org/en/cpi/2020/index/nor On Open Government in Norway, see, for example, Iris Nguyên Duy, The Norwegian

selected threats originating from foreign and/or national public or private actors: In the following, it is argued that the most relevant "digital threat" to access to information in order to form an informed opinion at the national level, is the imbalance resulting from wealth disparities and inequality of arms during the online political campaign (a.) while foreign threats to the electoral process and the electorate's freedom of speech have mostly taken the form of cyber-attacks and data collection, with a potential for disrupting both the parliamentary processes and elections (b.) in Norway.

### a. National Threats: The "unfairness" of Online Political Campaigns and the Consequences for Freedom of Expression

As most of the electoral campaign is currently taking place digitally (and to a lesser extent on TV and on the radio) this year, the digital platforms are playing a key role in the election campaign.[18] Both COVID-19 and the increased use of the internet in Norway, have contributed to a change in how the election campaign is conducted, and, to a large extent, in how the political actors reach the population and manipulate the voters, often with only a few people being aware of it. The technological change and digitalisation of the society opens for an effective dissemination of political information and direct communication between politicians and citizens. It is easier for resourceful political parties and candidates to send more political messages (resulting in an "outsized", disproportionate influence on the public debate), but also to better manipulate the electors, via *microtargeting* and because they can spend more for *political advertising on internet* and on the social media, and thus potentially reach more people and in a more sensible way (for their purpose). This has an impact on the freedom of expression of both the other, less wealthy political actors and of the electors, manipulated by targeted political ads, etc. And the ingrained "unfairness" of the bigger, wealthier political actors and their greater digital opportunities, leads to the weakening of the smaller actors' freedom to impart information: they do not

Experience with Open Government, in I. Bouhadana, W. Gilles, & I. Nguyên-Duy (eds.), *Parliaments in the Open Government Era*, Paris: Les Éditions IMODEV, pp. 1–32.

18 Kenneth Kandolf Haug, Valgkampen 2021: mindre dørbank, mer digitalt, *Dagsavisen*, 7 February 2021, https://www.dagsavisen.no/nyheter/innenriks/2021/02/07/valgkampen-2021-mindre-dorbank-mer-digitalt/ For the other consequences of digital campaigning, see section 3 a. See also section 3. A major change factor could be the infection and vaccination situation between June and September this year, as it will directly impact the way in which the political actors will be able/allowed to communicate with the electors (i.e., physically, via canvassing, campaign stands).

have the same means to disseminate their political program and ideas. There is a relative unfairness in media exposure, that goes beyond the "ordinary" one (i.e., the one a smaller party must expect because of its size and the number of supporters it has).

Anybody can post a political opinion on the internet nowadays for free. Political targeting is not illegal, nor is it necessarily problematic. It can ensure that voters will receive relevant information and increase civic engagement. But the strategy has changed: Recent studies have shown that *social media ads* have a significant effect on voters.[19] Political campaigning has been refined and has gone from aiming at wide target audiences to personalized influence through social media. And the new, digital, and technological, means and the extent to which political actors can target and influence the electorate are unsettling. The wealthier political parties (and their candidates) have theoretically the upper hand on the digital plan (even though a mainly digital campaign also gives smaller parties more visibility): Money can buy access to advanced technology (and support from external actors / consulting and IT companies) and to huge amounts of personal data, e.g., to create the psychographic profile of the data subjects and identify their location. The Cambridge Analytica data scandal in 2016 showed how those two factors could be exploited to become a powerful tool used for political influence and manipulation. They make it easier to micro-target key-people or groups of the population, and then produce tailor-made speech.[20] With the help of microtargeting, even "physical", door-to-door, canvassing is also more effective. One can also buy "visibility" and disseminate bigger amounts of speech more easily through online political advertising. This may constitute an interference with the other political parties and candidates' free speech (i.e., as they are relegated to the background / overshadowed by the bigger parties in the digital space). It also creates or increases a *de facto* inequality of arms, that results in the relative unfairness of online political campaigns (itself mainly due to differences in wealth) – and maybe the need to regulate this issue.

---

19 Federica Liberini, Michela Redoano, Antonio Russo, Angel Cuevas and Ruben Cuevas (2020), Politics in the Facebook Era - Evidence from the 2016 US Presidential Elections, CESifo Working Paper, No. 8235, Center for Economic Studies and ifo Institute (CESifo), Munich, https://www.econstor.eu/bitstream/10419/216631/1/cesifo1_wp8235.pdf

20 On the use of «dark ads» or «dark posts», see, for example, Issie Lapowsky, The Real Trouble With Trump's 'Dark Post' Facebook Ads, *Wired*, 20.09.17, https://www.wired.com/story/trump-dark-post-facebook-ads/

As a global trend, political parties are allocating increasingly larger shares of their marketing budgets toward digital platforms. Two attenuating key-factors are that Norway has a multi-party system (multiple political parties across the political spectrum run for national elections) and that the Norwegian political parties have relatively moderate election campaign budgets.[21] This limits their ability to purchase services that utilise micro-targeting of political messages. However, a recent report underlined the fact that targeting technology has become cheaper and more user friendly and that the current economic restrictions will not necessarily be an effective barrier against the widespread use of microtargeting technology in the near future, this year included.[22] While the second largest political party in Norway, the Labour party (*Arbeiderpartiet*) will prioritize contact with the electors via phone calls, the largest political party, the Conservative party (*Høyre*), announced in January that they would not use money for campaign in printed newspapers, and would prioritize both digital solutions (for example on Facebook and Snapchat) and canvassing instead.[23]

The use of these enhanced methods to manipulate the electorate, does not in itself inhibit free speech per se (everybody is at liberty to express themselves), but it causes an interference with the targeted part of the electorate, and it influences in a more invasive way their freedom of expression and opinion. Microtargeting "turns citizens into objects of manipulation and undermines the public sphere by

21 For the political parties budget for the last local and regional elections organised in 2019 and how much they used for political advertising on social media, see Øystein Tronsli Drabøs, Disse partiene bruker mest på valgkamp i sosiale medier, *NRK*, 7 September 2019, https://www.nrk.no/kultur/sa-mye-bruker-partiene-pa-sosiale-medier-1.14688027 and Martin Gundersen, Vi undersøkte Facebooks åpenhetsverktøy for å se hvordan politikerne vil vinne våre stemmer, *NRK Beta*, 5 September 2019, https://nrkbeta.no/2019/09/05/vi-undersokte-facebooks-apenhetsverktoy-for-a-se-hvordan-politikerne-vil-vinne-vare-stemmer/

22 Datatilsynet, Digital targeting of political messages in Norway, June 2019, https://www.datatilsynet.no/globalassets/global/dokumenter-pdfer-skjema-ol/rettigheter-og-plikter/rapporter/pa-parti-med-teknologien---engelsk.pdf The report also shows that Facebook is the central tool for the digital advertising of political messages, and the parties use mainly demographic data to target voters. See *ibidem*, p. 25.

23 Yngve Garel Svaldal, Høyre brukte over en halv million kroner på Snapchat-reklame i 2017 valgkampen – Her kan du se hva de skal gjøre i år, *KOM24*, 18 January 2021, https://www.kom24.no/erna-solberg-hoyre-rune-alstadsaeter/hoyre-brukte-over-en-halv-million-kroner-pa-snapchat-reklame-i-2017-valgkampen--her-kan-du-se-hva-de-skal-gjore-i-ar/282298

thwarting public deliberation, aggravating political polarization, and facilitating the spread of misinformation."[24]

*How are these Issues Regulated in Norway?*

Freedom of expression does not obligate the State to level the playing field during the electoral campaign, nor to "equalize resources through which individuals seek to influence public opinion or to guarantee equal access to the media to express one's views on matter of public concern."[25]

During the electoral campaign, television and radio broadcasters organize many electoral debates, to which all major parties (those with a vote share larger than 3 % in the previous election) have fair access. However, there is no direct government interference in choosing the teams of journalists that conduct debates, and, in general, representatives of the larger political parties are interviewed more often and participate in more debates than do small-party representatives.

Candidates and parties are free to purchase political advertising in print publications and on the internet. There is however a total ban on political advertising on linear TV (and teletext) in Norway, because it may give an unfair marketing advantage to the wealthiest political actors, and because it may also give a distorted picture of complex societal issues, in addition to the fact that the form of advertising does not provide much opportunity to respond.[26] This ban was criticised by the European Court of Human Rights in December 2008, in the case of the Rogaland Pensioners' Party.[27] This small political party had received almost no TV coverage and found it difficult to make people aware of its message. They had purchased television advertising from TV Vest AS, a local television station. However, section 3-1 (3) of the *Broadcasting Act 1992*

---

24 W. Gorton, 'Manipulating Citizens: How Political Campaigns' Use of Behavioral Social Science Harms Democracy', (2016) 38 *New Political Science*, no. 1, https://doi.org/10.1080/07393148.2015.1125119, pp. 61–80, p. 62. See however Pablo Barberá, Social Media, Echo Chambers, and Political Polarization, in Nathaniel Persily, Joshua A. Tucker (eds.), *Social Media and Democracy*, Cambridge University Press, 2020, chap. 3, pp. 34–55, https://doi.org/10.1017/9781108890960

25 Ashutosh Bhagwat & James Weinstein, Freedom of Expression and Democracy, *op. cit.*, p. 89 note 35.

26 Medietilsynet, Forbud mot politisk reklame på tv, 30.08.19, https://www.medietilsynet.no/om/aktuelt-2019/forbud-mot-politisk-reklame-pa-tv/

27 ECHR, *TV Vest As and Rogaland Pensjonistparti v. Norway*, 11 December 2008, application no. 21132/05.

read: “Broadcasters cannot transmit advertisements for life philosophy or political opinions through television”.[28] The case concerned a fine imposed on TV Vest for showing adverts for the Pensioners Party without authorisation, prior to the local and regional elections of 2003. TV Vest appealed to the Oslo City Court and then to the Supreme Court, which both upheld the State Media Authority’s decision to impose the fine. TV Vest then appealed to the European Court of Human Rights. The Norwegian government argued that the purpose with the ban “was to guarantee political expression by ensuring fairness and equality, as well as preserving the quality of political debate. (…) The possibility of advertising on television would clearly benefit the wealthier and/or established interests in society. There was thus a need to avoid de facto discrimination, distorting democratic processes in favour of the wealthy and powerful. (…) The prohibition ensured the political impartiality of television broadcasting. It also had the effect of limiting the total amount of money spent on election campaigns by political parties and interest groups, reducing their dependence on wealthy donors and ensuring a level playing field in elections.”[29] The Court held that there was not a reasonable relationship of proportionality between the legitimate aim pursued by the prohibition on political advertising and the means deployed to achieve that aim. The restriction (*i.e.*, the prohibition of political advertising on TV and the fine imposed on TV Vest) could therefore not be regarded as having been necessary in a democratic society (as demonstrated by the different restrictions imposed in other European countries), within the meaning of art. 10 (2) ECHR for the protection the rights of others – in other words, that there was a violation of art. 10 ECHR. Yet, the Norwegian government did not interpret the ECHR’s decision as requiring them to amend the legislation and did not lift the ban.[30]

Rather paradoxically, political advertisement is allowed on all other kinds of media such as the radio, podcast, internet, and social media.[31] That is where the

---

28 In addition, political parties may not sponsor programs on television, radio or in audio-visual ordering services, according to the Broadcasting Regulations § 3–11 (2). Political parties cannot buy product placement in these channels either, according to the Broadcasting Act § 3–7 f.

29 *TV Vest As and Rogaland Pensjonistparti v. Norway*, §§ 43–44.

30 For the minimum changes made by the Government to comply with the ECtHR’s decision, see St.Meld. nr. 18 (2008–2009). For a critic, see Fredrik Sejerstad, Rettsliggjøring og reklame. En studie i europeisk rettsliggjøring og nasjonalt handlingsrom, in *Politikk i grenseland- festskrift til Øyvind Østerud*, Universitetsforlaget, Oslo, 2014, p. 267.

31 Medietilsynet, Forbud mot politisk reklame på tv, 30.08.19, *NTB.no*, https://kommunikasjon.ntb.no/pressemelding/forbud-mot-politisk-reklame-pa-tv?publisherId=89456&releaseId=17869896 (last accessed: 02.08.21).

actual problem (or a contemporary threat) lies, at a time when digital communication reigns over all other communication channels, especially in times of pandemic: With no regulation of political advertising online, the threshold for manipulation is low – and the risks for the electors' freedom of expression higher.

The *General Data Protection Regulation* (GDPR), implemented in Norway in 2018 (see the *Personal Data Act* of 15 June 2018), sets the framework for how personal data can be used, but is not enough alone. *Digital ethics*, for example in the form of implementation of codes of conduct in the digital space, are still underdeveloped in Norway. However, the Norwegian Board of Technology suggested in 2018 to develop a "code of conduct for digital election campaigns" along the following guidelines:

- Openness and labelling
  The political parties must be open about how they use data and place labels on online- and social media ads. The labelling can include information about how much money has been used, and who is in the target group. Political ads should also be stored in an archive.
- A budgetary rule for data and money
  The parties should in unison determine what types of data can be used for targeted political advertisements. They should also decide whether there needs to be introduced a spending limit for digital marketing.
- Campaign audits
  The parties must disclose how much money was spent on digital marketing and how this money was spent. Data audits should be introduced as a way of ensuring that data is only used for its intended purpose.[32]

It has been partially implemented in 2019, and now in 2021:

To prevent voters from being manipulated ahead of elections and to increase transparency, the Norwegian government is cooperating with big social media platforms like Facebook. But nothing is regulated, the available information on the collaboration itself is scarce, and the scope and impact of this collaboration are currently rather limited. Facebook has for example set up an "Ad library", giving access to all to ads about social issues, elections or politics, including spend, reach and funding entities, thus increasing transparency around the election campaign on the platform and theoretically helping in preventing

32 Tore Tennøe, Marianne Barland, Elections, technology and political influencing, *Teknologirådet*, 2019, p. 2, https://teknologiradet.no/wp-content/uploads/sites/105/2019/06/Elections-technology-and-political-influencing.pdf

interference in elections.[33] But one of the biggest flaws of this measure is that it all depends on what information or data the political actors are willing to share.

The political parties have not yet come to a common agreement on the types of data that can be used for targeted political advertisement. The need to regulate the use of microtargeting, especially as a tool during the election campaign, should be assessed by the authorities. But, as indicated earlier, the use of microtargeting seems to be still limited in Norway, because of the political parties financial and human resources,[34] the prevalent concept of *trust* in the Norwegian society and the GDPR, that has been hanging over the political parties like a Sword of Damocles, with the risk of heavy fines, since 2019.

There is no legislation on unlimited independent spending in Norway. There are no limits on contributions and expenditures used by individuals or groups to advocate candidates or causes in the public arena. There is however a system for digital reporting and disclosure on political parties' finances. The parties are "required to know who the donor is, that it is not a state-owned enterprise or a foreign donor. In connection with election years, the parties have a special duty to report donations exceeding NOK 10 000, received in the period 1 January until the Friday before the election day."[35] The Norwegian political parties spent approximatively NOK 95 million (9,5 M €) during the last parliamentary election campaign in 2017 (more than 50 % of the total sum was spent by the two largest parties).[36]

---

33 Facebook, About the Ad Library, https://en-gb.facebook.com/business/help/2405092116183307?id=288762101909005 Until January 2021, Twitter also had an "Ads Transparency Center". Andrew Hutchinson, Twitter Quietly Removes its Ads Transparency Center, 3 February 2021, *Social Media Today*, https://www.socialmediatoday.com/news/twitter-quietly-removes-its-ads-transparency-center/594485/

34 However, it will be interesting to compare the numbers and statistics in the aftermath of the parliamentary election. There might be some surprising results.

35 On election campaign contribution in Norway, see https://www.partifinansiering.no/en/election-campaign-contributions/ (last accessed: 05.05.21).

36 Espen Breivik, Gaute Zakariassen, Bruker 95 millioner på årets valgkamp, *NRK*, https://www.nrk.no/norge/bruker-95-millioner-pa-arets-valgkamp-1.13670933; For statistics on the political parties' revenue i 2017, see SSB, Inntektene opp 43 prosent siden 2005, 4 September 2017, https://www.ssb.no/valg/artikler-og-publikasjoner/inntektene-opp-43-prosent-siden-2005

## b. The Informed Electorate and Enhanced Digital Manipulation of Information by Foreign Actors

Freedom of political expression includes the expression of information and ideas through which the speaker participates in the public debate and tries to influence public opinion. It is vital for a thriving democracy that there is an unrestrained flow of political information.[37] In the same spirit, freedom of opinion needs to be absolute: Any individual should have the right to hold an opinion without interference. This should not be tempered with, by, for example, influencing the conscious or subconscious mind by using manipulation, brainwashing, echo chambers, epistemic/filter bubbles, etc., that pervert the flow of information and alter our perception of the truth and, ultimately, the forming of our (informed) opinions/minds.[38]

Again, tentative to manipulate information is not a new issue[39] – and not necessary illegal in itself.[40] What is challenging everywhere (not just in Norway) is

37 "freedom of expression is one of the 'conditions' necessary to 'ensure the free expression of the opinion of the people in the choice of the legislature'. For this reason, it is particularly important in the period preceding an election that opinions and information of all kinds are permitted to circulate freely." – See ECtHR, *Bowman v. United Kingdom* (1998) 26 EHRR 1, [42] (citations omitted).

38 C Thi Nguyen, Escape the echo chamber, *Aeon*, 9 April 2018, https://aeon.co/essays/why-its-as-hard-to-escape-an-echo-chamber-as-it-is-to-flee-a-cult (last accessed: 28.04.21).

39 "For example, the U.S. and the USSR/Russia have intervened in one of every nine competitive national level executive elections between 1946 and 2000." – Dov H. Levin, When the Great Power Gets a Vote: The Effects of Great Power Electoral Interventions on Election Results, *International Studies Quarterly*, Vol. 60. I. 2, 2016, pp. 189–202, https://academic.oup.com/isq/article/60/2/189/1750842

40 It is quite interesting that the part of Section 105 of the previous Penal Code of 1902, on liability for seeking to influence another person's behaviour or voting in public affairs or to prevent another person from voting «by false inducements» (*løgnagtige Forespeilinger*), was not kept in the revised version of the Penal Code. Section 105 (now section 151 of the new Penal Code) aimed at protecting the citizens' right to participate in elections in a free and independent manner. The provision was not specifically aimed at politicians but applied to everyone. Truly enough, it was considered as a sleeping paragraph, it was probably never intended to be used to prosecute politicians who lied and did not fulfil their election promises, and the removed part "by false inducements" was problematic on many levels. But it also means that, since 2015, it is longer punishable to lie to the electors or to make false promises one never intended to fulfil in Norway. See Øyvind Bye Skille, Endret politikerne loven slik at

the speed, scope and scale of misinformation (the inadvertent sharing of false information) and disinformation (the deliberate creation and sharing of information known to be false) enabled by technology, and the difficulty to even canalize the digital manipulation of information (especially when incredibly advanced tools such as deep fakes are used), because of content generated or manipulated by AI / automated bots, bot-networks, troll factories, and even our own sharing-activities online.[41]

The issue is all the more acute that Norway has an excellent communication infrastructure: According to the information service MediaNorway, 98 % of Norway's population had already access to the Internet in 2019, and 95 % had access to a smartphone, which means that most people carry a computer that is connected to the Internet and which has sensors such as a camera, GPS and a motion sensor[42] – and are reachable by (*i.e.*, at the mercy of) those who have sufficient knowledge in the new technologies. Many Norwegians are also very active on the digital and social media. Norway is also a big "media nation": with a little more than 5 million inhabitants, it has a dozen TV channels, hundreds of radio channels and more than 300 local and national newspapers.[43] A recent survey (2021) has revealed that 4 out of 10 Norwegians have come across fake news in 2020–2021 and 7 out of 10 over information they doubted was true.[44] 5 out of 10 Norwegians have also come across fake news about the COVID-19. Dissemination of disinformation and fake news seems to be easier online, on

de kan lyve til velgerne?, *Faktisk.no*, 30.09.20, https://www.faktisk.no/artikler/YVN/endret-politikerne-loven-slik-at-de-kan-lyve

41 Definitions of misinformation and disinformation by Claire Wardle. See Claire Wardle, Fake news. It's complicated., FirstDraft, 16 February 2017, https://firstdraftnews.org/latest/fake-news-complicated/ (last accessed: 28.04.21).

42 This applies to people aged from 9 to 79 years old. See the statistics on https://medienorge.uib.no/statistikk/medium/ikt/347 and https://medienorge.uib.no/statistikk/medium/ikt/388; See also Teknologirådet, This Time it's Personal. The Digital Shift in the Public Sector, 2018, https://teknologiradet.no/wp-content/uploads/sites/105/2018/05/Report_This-time-its-personal_The-digital-shift-in-the-public-sector.pdf

43 Webinar: Falske nyheter og desinformasjon – hvor utbredt er det?, 18 February 2021, https://www.youtube.com/watch?v=FkpGCgytkEA (last accessed: 19.04.21).

44 Medietilsynet, Undersøkelse om kritisk medieforståelse i den norske befolkningen. Delrapport 1: Falske nyheter og desinformasjon, February 2021, https://www.medietilsynet.no/globalassets/dokumenter/rapporter/kritisk-medieforstaelse-2021/210218-kmf-delrapport-1-falske-nyheter.pdf (last accessed 19.04.21).

the Internet, and the media of choice seems to be Facebook (1 of 3 has seen fake news on COVID-19 on Facebook).[45]

Moreover, the contemporary exercise of freedom of opinion and expression owes much to private industry, in particular to big tech companies and social media platforms, which act as a gateway for information and an intermediary for expression: They play a key role in the free flow of information, ideas and opinions. They enable the public and individuals to participate actively in the communication process. But it is precisely through this same gateway that malicious actors may influence the individuals.[46]

In that respect, foreign intelligence operations and influence seem to represent some of the greatest digital threats Norway faces in 2021.[47] These digital threats may take the form of intelligence operations, data theft and disinformation / information manipulation campaigns (when foreign actors do not simply buy influence online), separately or combined. They threaten the digital information ecosystem and the digital public sphere. They test the resilience of the democratic system and impact, more or less directly, freedom of expression.

Although it is a small country, Norway is an attractive intelligence target because it is geo-strategically important, it has large natural resources, and it

---

45 *Ibidem.*

46 Big social media platforms such as Facebook, Google (who owns YouTube) and Twitter are expanding their domination on media distribution. They have become a place of choice for public debate. But they also keep using algorithms that gather data / information from the users and that contribute to, *inter alia*, this information issue that has direct consequences for the forming of opinions and thus for the freedom of expression. They keep "a social-validation feedback loop (…) exploiting a vulnerability in human psychology" (Sean Parker) to make profit. – See Olivia Solon, Ex-Facebook president Sean Parker: site made to exploit human 'vulnerability', *The Guardian*, 9.11.17, https://www.theguardian.com/technology/2017/nov/09/facebook-sean-parker-vulnerability-brain-psychology. They rely heavily on dopamine addiction strategy. – Simon Parkin, Has dopamine got us hooked on tech?, *The Guardian*, 4.03.18, https://www.theguardian.com/technology/2018/mar/04/has-dopamine-got-us-hooked-on-tech-facebook-apps-addiction. By personalising and filtering their content, based on the data they have gathered on the users, they participate in generating echo chambers and filter bubbles and contribute to the polarisation of the public debate. They thus also form a private and international threat to freedom of speech, but it is latent, pervasive, and not particularly specific to an electoral context. This paper focuses on digital national and foreign threats in the specific context of elections only.

47 But disinformation, information manipulation may of course also come from national or local groups or individuals, such as experienced during the local elections in Norway in 2019. (Kristoffer Egeberg, interview)

develops world-leading technology in many areas.[48] And because of the need to deal with the pandemic at the same time, it has also become more vulnerable.

Both the Norwegian Police Security Service (PST) and the Norwegian Intelligence Service's threat assessments for 2020 and 2021 have highlighted Norway's digital vulnerabilities and the intensification of foreign influence activity against democratic processes and public opinion in Norway. They show that state-controlled espionage in the digital space (with Russia and China at the forefront[49]) represents a persistent and serious threat to Norway, and they worry that these will have a negative impact on the 2021 elections.[50] Their fear was confirmed last autumn and in March this year.[51]

---

48 Geir Hågen Karlsen, Dataangrepet: Hva er de ute etter?, *Dagens næringsliv*, 07.09.20, https://www.dn.no/innlegg/cyberangrep/datasikkerhet/pst/innlegg-dataangrepet-hva-er-de-ute-etter/2-1-869074 There are ongoing strategic tensions between Russia and Norway in the Artic due to a number of military and economic reasons.

49 Nora Buli, Norway says cyber attack on parliament carried out from China, *Reuters*, 19.07.21, https://www.reuters.com/world/china/norway-says-march-cyber-attack-parliament-carried-out-china-2021-07-19/. With their spying operation on France, Germany, Norway and Sweden in May 2021, the United States may now be added to the list. Reuters, U.S. spied on Merkel and other Europeans through Danish cables - broadcaster DR, *Reuters*, 30.05.21, https://www.reuters.com/world/europe/us-security-agency-spied-merkel-other-top-european-officials-through-danish-2021-05-30/ (last accessed: 30.07.21).

50 Norwegian Intelligence Service, *Focus 2021*, 2021, https://www.forsvaret.no/aktuelt-og-presse/publikasjoner/fokus/rapporter/Focus2021-english.pdf/_/attachment/inline/450b1ed0-1983-4e6b-bc65-4aa7631aa36f:21c5241a06c489fa1608472c3c8ab855c0ac3511/Focus2021-english.pdf (last accessed: 28.04.21), pp. 6, 14–23. See also Nasjonal sikkerhetsmyndighet, *Risiko 2021*, p. 7–14, https://nsm.no/getfile.php/136419-1616673370/Demo/Dokumenter/Rapporter/NSM_Risiko_2021_web_enkeltside_1203.pdf. According to the Norwegian Directorate for Civil Protection (DSB), the Norwegians are concerned by/about cyber-attacks. 39 % of the population is worried that cyber-attacks will knock out important management systems over the next five years. DSB, Befolkningsundersøkelsen 2020, 2020, https://www.dsb.no/globalassets/dokumenter/rapporter/andre-rapporter/rapport---befolkningsundersokelse-om-husholdningers-egenberedskap-2020.pdf

51 After the second cyberattack, in March 2021, the President of the Storting, Tone Wilhelmsen Troen, declared that it was an attack on democracy and that its "severity is underscored by the fact that this is happening in the run-up to a parliamentary election and as parliament is handling a pandemic". – *Reuters*, Norway's Parliament Hit by New Hack Attack, *US News*, 10 March 2021, https://www.usnews.com/news/world/articles/2021-03-10/norway-parliament-sustains-fresh-cyber-attack

The "large-scale" cyberattacks against the Storting in the autumn of 2020 (allowing the perpetrators to break into the emails of some parliamentarians and staff) and in March 2021 had probably state support,[52] and clearly showed both that the digital intelligence threats against the Norwegian democratic institutions are very real, and that Norway is not well enough prepared to tackle them. The stolen data may be used either against the government or to pressure selected/ targeted persons and get them to do what they want. But the hacking attack itself also creates tensions and divisions in the population. Indeed, the main strategy to manipulate the mass mind is to create and maintain a climate of fear and distrust (or even just doubt in people's mind), and then capitalize on declining levels of trust in institutions and experts, influence the result of opinions of the voters this way or with covert propaganda, and thus influence the result of the elections in a country that is not theirs.[53]

Disinformation may be defined as "false, inaccurate, or misleading information designed, presented and promoted to intentionally cause public harm or for profit."[54] It "includes forms of speech that fall outside already illegal forms of speech, notably defamation, hate speech, incitement to violence, etc. but can nonetheless be harmful."[55] The use of disinformation has escalated during the covid-19 pandemic.[56] We are focusing here of a high-risk form of

52 «In autumn 2020 a number of cyber attacks were carried out against the Storting and certain Norwegian organisations. PST's investigation revealed that the cyber espionage group Fancy Bear, also known as APT-28, was probably behind the attacks. The group is part of the Russian military intelligence service GRU». – The Norwegian Police Security Service (PST), *National Threat Assessment 2021*, https://www.pst.no/globalassets/artikler/trusselvurderinger/nasjonal-trusselvurdering-2021/ntv_2021_eng_web_180221.pdf A massive amount of data from a broad spectrum of politicians was stolen.

53 For other methods to undermine the conduct of the election, see, for example, Geir Hågen Karlsen, Slik kan fremmede makter undergrave stortingsvalget, *Dagens næringsliv*, 10.02.21, https://www.dn.no/innlegg/forsvaret/teknologi/digitalisering/innlegg-slik-kan-fremmede-makter-undergrave-stortingsvalget/2-1-960364

54 European Commission, A multi-dimensional approach to disinformation, Report of the independent High-level Group on fake news and online disinformation, 2018, p. 5.

55 Ibidem, p. 10.

56 The Norwegian Media Authority has revealed that 45 % of the Norwegians had come across fake news about the Covid 19-eruption and that 38 % had seen fake news on social media during a one week-period in March 2020. – Medietilsynet, Undersøkelse om falske nyheter i den norske befolkningen, 2020, https://www.medietilsynet.no/globalassets/publikasjoner/2020/203030-_omnibus-om-falske-nyheter-i-forbindelse-med-korona-fra-medietilsynet.pdf

disinformation,[57] where foreign actors try to undermine the Norwegian political process and values, such as democracy and freedom of expression. There is growing evidence of the extent and degree of sophistication of manipulation of technology platforms; and even though it is difficult to determine the concrete effect of such informational manipulation operations, it has been shown that their reach is significant and cannot be overseen.[58] Regarding their impact on freedom of expression, the manipulation of information and the campaigns that are used to sow distrust in the good functioning of the institutions and of democracy nearly amounts to a new type of (insidious) censorship strategy:[59] How can one form an opinion based on facts, when one is no longer sure whether the information is correct?

Truly enough, in its report on foreign information manipulation during the local elections of 2019, the independent research institute SINTEF did not find any clear signs of foreign influence, but it should be underlined that they only considered local elections, not national ones:[60] There might be less interest for

---

57 As opposed to "low-risk forms such as honest mistakes made by reporters, partisan political discourse, and the use of click bait headlines". – See European Commission, A multi-dimensional approach to disinformation, Report of the independent High-level Group on fake news and online disinformation, 2018, p. 10.

58 See Disinformation and propaganda – impact on the functioning of the rule of law in the EU and its Member States, Study, 2019, https://www.europarl.europa.eu/RegData/etudes/STUD/2019/608864/IPOL_STU(2019)608864_EN.pdf

59 Associate Professor and techno-sociologist Zeynep Tufekci explains that new censorship strategies often include the spread of disinformation, bots with intentions to troll and distract, and harassment campaigns. "The most effective forms of censorship today involve meddling with trust and attention, not muzzling speech," Tufekci writes. "These tactics usually don't break any laws or set off any First Amendment alarm bells. But they all serve the same purpose that the old forms of censorship did: They are the best available tools to stop ideas from spreading and gaining purchase." – Zeynep Tufekci, It's the (Democracy-Poisoning) Golden Age of Free Speech, *The Wire*, 16 January 2018.

60 Tor Olav Grøtan, Jannicke Fiskvik, Peter Halland Haro, Per Gunnar Auran, Bjørn Magnus Mathisen, Geir Hågen Karlsen, Melanie Magin, Petter Bae Brandtzæg, *På leting etter utenlandsk informasjonspåvirkning. En analyse av det norske kommunestyre- og fylkestingsvalget 2019*, SINTEF, 2019, https://www.regjeringen.no/contentassets/4d850821991746ecbcd9477a475baf73/sintef-rapport_2019-01292_gradering_apen.pdf SINTEF also confirmed in the report that hidden networks and fake news on social media pose a challenge to democracy. See also Tor Olav Grøtan, Jannicke Fiskvik, Peter Halland Haro, Petter Bae Brandtzæg, Mistenkelige debattanter sprer desinformasjon, skjuler sine spor, *Dagens næringsliv*, 26.02.20, https://www.dn.no/innlegg/sosiale-medier/medier/fake-news/innlegg-mistenkelige-debattanter-sprer-desinformasjon-skjuler-sine-spor/2-1-761058?fbclid=IwAR3A0hqvnxE1FwqoqlocZSfL3hbCIhRWuKIEhvHwlJi3dcbfSK6X_7hbY4Y

foreign actors in manipulating local elections rather than national ones. The Intelligence services reports and the two cyber-attacks on the parliament tend to prove that the abovementioned threats are real.

## *What can be Done to Counter these Threats? How are they Legally Tackled in Norway?*

It must always be borne in mind that the introduction of measures to increase security in elections can potentially conflict with safeguarding other important values and principles such as freedom of expression. The legal obligation imposed by Article 2 ICCPR is that the State must respect, protect, and refrain from interference in these rights either by the State organs itself or by third parties (see also article 19 ICCPR and art. 10 (2) ECHR, both incorporated in the Human Rights Act).

Ten (non-regulatory) measures were announced by the Government in 2019.[61] Several measures have already been taken, such as the establishment of a national Cyber Security Centre, the NCSC, whose mission is to enhance Norway's resilience in the digital domain (it is part of Norwegian National Security Authority (NSM)). The government has also implemented a national strategy to ensure access to digital security expertise and allocated NOK 800 million for the purpose, as well as NOK 500 million for a technical security package to strengthen the ability to detect and handle attacks, and a new *Intelligence Service Act* was adopted in 2020 to give the Intelligence Service significantly "better opportunities" to expose intelligence activities and digital attacks on Norway.[62]

Sections 151 to 154 of the Penal Code already address the issue of influence over voting and over election results to a large extent.[63] The whole chapter 21 of

61 Regjeringen, Ti tiltak for hindre uønsket påvirkning i valggjennomføringen, 21.06.19, https://www.regjeringen.no/no/aktuelt/ti-tiltak-for-hindre-uonsket-pavirkning-i-valggjennomforingen/id2661220/

62 Ny e-lov gir bedre beskyttelse mot digitale angrep, Discourse by the Minister of Defence Franck Bakke Jensen, 14.09.20, https://www.regjeringen.no/no/aktuelt/elovenbeskyt ter/id2741266/ (last accessed: 23.04.21).

63 The previous Penal Code of 1902 (valid until 2015) had a provision on disinformation in its chapter 8 on felonies against the independence and security of the State, section 97b: "Any person who against his better judgment or through gross negligence publicly spreads or reports to a foreign power false rumours or incorrect information which if they gain credence are likely to endanger the internal and external security of the realm, or any relationship to foreign powers, or who aids and abets thereto, shall be

the Penal Code deals with protection of information and exchange of information. Intrusion into a computer system, hacking included, by breach of a protective measure or by other illicit means, constitutes a criminal offence according to article 205 of the Penal Code, while article 206 sanctions denial-of-service attacks and infection of IT systems with malware.[64] And article 5 of the same act sets forth the condition for its extra-territorial application. But there are no specific legal provisions sanctioning hacking and attempted hacking of electoral systems.

In addition, the consulting company Proactima AS submitted its report on "Safety in democratic processes in Norway" to the Electoral Law Committee in 2019 and made 9 regulatory proposals, as well as 20 other possible measures, while pointing out the challenges they constitute for openness and freedom of expression.[65] Among others: to establish a contingency provision in the election legislation, establish / clarify legal basis for sanctioning the hacking, and attempted hacking, of electoral systems, establish a legal basis for the supervision of safety regulations for the election, etc.

In her book on fake news, Associate Professor in Communication Bente Kalsnes recommends measures in four areas: media, technology and platforms, legislation, as well as education of children and adults in critical investigation of sources and in media and information literacy.[66] But she only briefly lists up the legislative measures taken by Germany, France, Italy and Ireland against fake news in the "regulatory part" of her proposals, without either criticising them or adapting them to the Norwegian case.

Lieutenant Colonel Geir Hågen Karlsen makes 38 proposals to fight foreign influence in several articles this year, some of them potentially challenging both

liable for imprisonment for a term not exceeding two years." – Norwegian Ministry of Justice, The General Civil Penal Code, unofficial translation, 2006, https://app.uio.no/ub/ujur/oversatte-lover/data/lov-19020522-010-eng.pdf . The legislator did not keep the provision for the new Penal Code of 2005. For an explanation, see *inter alia* Ot.prp. nr. 8 (2007–2008), at 7.3.2.3 and 7.12.2.7, https://www.regjeringen.no/no/dokumenter/otprp-nr-8-2007-2008-/id488010/sec7

64 See Stian Oddbjørnsen et alia, Norway: Cybersecurity Laws and Regulations 2021, *ICLG.com*, 02.11.20, https://iclg.com/practice-areas/cybersecurity-laws-and-regulations/norway (last accessed: 07.05.21).

65 Anne-Kari Valdal, Hermann S. Wiencke, Chris Dale, Svein Tuastad, Trine Holo, Willy Røed, Bjørg Sandal, Sikkerheten i demokratiske prosesser i Norge, in NOU 2020: 6, Annex 6, pp. 478–512. The 9 regulatory measures are summed up p. 479.

66 Bente Kalsnes, Falske nyheter. Løgn, desinformasjon og propaganda i den digitale offentligheten, Cappelen Damm Akademisk, 2019.

democracy and freedom of expression. He does not elaborate on the regulatory proposal, but he underlines the importance of ensuring "a fair balance between threats, measures and consequences, and prevent measures to protect democracy itself from undermining democracy".[67] Any regulation of the contemporary threats to freedom of expression may entail direct and indirect threats to the same freedom and even other rights – and should thus be well thought through and carefully framed. The abovementioned regulatory proposals show a certain incapacity to find satisfying legal solutions – or to see how to find legal solutions that do not entail an infringement of one or several human rights.

Lastly, the Norwegian fact-check website *faktisk.no* evaluates news articles and statements, in order to disclose misinformation, disinformation and fake news. *Faktisk.no* was created in 2017 by two of Norway's largest newspapers and by the Norwegian Broadcasting Corporation (NRK), the Norwegian government-owned radio and television public broadcasting company, and the largest media organisation in Norway. But their resources are limited. One should consider creating several local branches of fakta.no to support and extend their action, or even new fact-check websites, nationwide. Several media, companies, universities and research facilities have also recently developed a Norwegian Media Cluster that might contribute to the fight against media manipulation.[68]

## 2. Online Threats and Digital Hate Speech: Their Chilling Effect on Freedom of Speech

Digitalization has *inter alia* entailed a democratization of the possibility to speak publicly. With the development of the media and new technologies, most people can easily create and share content with others, participate in the public debate, and they usually do so online, especially on the social media. But the same digitalization has also entailed a blurring of the conditions for boundary-drawing related to free speech. In *Delfi AS v. Estonia*, the ECHR has stated that "[d]efamatory and other types of clearly unlawful speech, including hate speech and speech inciting violence, can be disseminated like never before, worldwide,

67 Geir Hågen Karlsen, Slik kan vi beskytte stortingsvalget mot fremmed påvirkning, *Dagens næringsliv*, 19.03.21, https://www.dn.no/forskningviser-at-/stortingsvalget-2021/dataangrep/russland/innlegg-slik-kan-vi-beskytte-stortingsvalget-mot-fremmed-pavirkning/2-1-983629; See the proposed measures in Geir Hågen Karlsen, Hvordan kan vi beskytte valg mot fremmed påvirkning?, *Internasjonal Politikk*, vol. 79–1 (2021), pp. 90–113, at p. 100–7, https://doi.org/10.23865/intpol.v79.2309;

68 https://mediacitybergen.no/about/

in a matter of seconds, and sometimes remain persistently available online."[69] It has also underlined that "the risk of harm posed by content and communications on the Internet to the exercise and enjoyment of human rights and freedoms, particularly the right to respect for private life, is certainly higher than that posed by the press".[70]

In times of elections, there is an even more pressing need to be able to take part in the public debate for *both* the electors and the candidates, and even the general public. In this context, misinformation, disinformation, and malevolent content in the form of threats and hate speech may cause political instability in society and disrupt the general debate. A political election is not a licence for offence or defamation.[71] But in Norway, a main issue is the increase in threats and hate speech online. This section will outline this issue and present the current legal framework.

## a. A Shift in the Perception of Threats and Hate Speech Online:

While a research report from 1990 concluded that the Norwegian population was not racist,[72] a report from 2019 shows a rather different result thirty years later.[73] In the news media, 1 out of 4 Norwegians is presented as being

---

69 ECHR [GC], *Delfi AS v. Estonia*, Application no. 64569/09, 16 June 2015, § 110.

70 Ibidem, § 133.

71 I will rule out the issue of *defamation* for the purpose of this article: defamation of public debaters/speakers and of political candidates to damage their good reputation and reduce their changes at the election, is usually made on the national or local level, seldom on an international one. Defamation (and insults to government institutions and officials) is no longer a criminal offence in Norway. It was decriminalised with the amendment of the Penal Code of 2005 (effective from 1 October 2015). However, defamation can in certain cases be penalised as a violation of privacy under Section 267 of the Penal Code, and the Damage Compensation Act of 13 June 1969 (Sections 3 to 6) provide a legal basis for claiming compensation due to defamatory content, in civil cases. See Advokatfirmaet Thommessen, Digital content regulation and management in Norway, 14 January 2019, https://www.lexology.com/library/detail.aspx?g=a9395fce-8cf4-4a92-a7c5-8ec3c6318fd5 (last accessed: 21.04.21).

72 Hernes, G. & Knudsen, K. (1990). Svart på hvitt. Norske reaksjoner på flyktninger, asylsøkere og innvandrere, Oslo: Fafo-rapport, 1990: 109, p. 132, https://www.fafo.no/zoo-publikasjoner/fafo-rapporter/item/svart-pa-hvitt

73 Tyldum, G. (2019). Holdninger til diskriminering, likestilling og hatprat i Norge. 2. utgave, Oslo: Fafo rapport 2019: 26, https://www.fafo.no/zoo-publikasjoner/summaries/item/norwegians-attitudes-towards-gender-equality-hate-speech-and-the-instruments-of-gender-equality-policy-2nd-edition

"racist".[74] Particularly in the online space, hate speech has become a widely recognised societal problem.[75] "In social media, users can express hateful thoughts and attitudes without a tangible contact with victims, anonymously, and relatively free from external control (…)."[76] In Norway, threats and hate speech (especially on social media) have been one of the most discussed topics these past two years. "Allegations of threats and incitement are many, and they come from individuals, from PST, from the Police, from the Norwegian Association of Local and Regional Authorities, the President of the Storting, and the Art community to name a few", recognised the leader of the current Norwegian Commission for Freedom of Expression, tasked with studying the social, technological, legal, and economic frameworks for freedom of expression in Norway today.[77] The Supreme Court's recent case-law also reflects the extent of this social issue.

*Threats*[78] (as opposed to expressions of opposition and of displeasure in the public political discourse) and hate speech can silence political debaters, and even lead politicians to quit politics, both because they are intimidated and/ or because they lose their commitment and faith in the police, the courts, the institutions, and in the democratic system. Several politicians (especially women with a minority background),[79] and many of the survivors of the 2011 Norway

74 Julia Orupabo, Spranget fra hvem som er rasist, til når, hvor og hvordan, *Tidsskrift for samfunnsforskning*, 01 / 2021 (Volum 62), pp. 116–120, p. 116, https://doi.org/10.18261/issn.1504-291X-2021-01-10

75 Fischer, A., Halperin, E., Canetti, D., & Jasini, A. (2018). Why we hate, *Emotion Review*, 10(4), 309–320, p. 315, https://doi.org/10.1177/1754073917751229

76 Kaakinen, M., Oksanen, A., Gadarian, S.K., Solheim, Ø.B., Herreros, F., Winsvold, M.S., Enjolras, B., & Steen-Johnsen, K. (2021), Online Hate and Zeitgeist of Fear: A Five-Country Longitudinal Analysis of Hate Exposure and Fear of Terrorism After the Paris Terrorist Attacks in 2015, *Political Psychology*, p. 4, https://doi.org/10.1111/pops.12732

77 Kjersti Løken Stavrum, Nei, samfunnsdebatten er ikke full av hat, *VG*, 22.04.21, https://www.vg.no/nyheter/meninger/i/vAJzaL/nei-samfunnsdebatten-er-ikke-full-av-hat

78 Storting politicians are protected by sections 115 and 116 of the Penal Code on threats against, among others, ministers and Storting politicians. The Police Security Service (PST) is responsible for their safety and security. Due to the increase of hate speech in the recent years, members of the youth parties have also been placed under the PST's protection.

79 See, for example, Martin Lægland, Ida Lyngstad Wernø, Caisa Linea Hagfors, Oppfordret til grov vold mot Lan Marie Berg: – Jeg vil beklage, *VG*, 24.05.21, https://www.vg.no/nyheter/innenriks/i/AleEqz/oppfordret-til-grov-vold-mot-lan-marie-berg-jeg-vil-beklage (last accessed: 26.05.21).

attacks,[80] as well as several journalists,[81] police officers and employees in the Child Welfare services, have been, and are still being, the victims of violent threats (sometimes death threats) and hate speech by (anonymous or not) Internet trolls. The Internet trolls usually target people because of their differences (race/ ethnicity, religion, sexual preferences, gender, disabilities, etc.) and/or their professional occupations. The police and the prosecutors have gradually taken hold of the matter and reacted actively to digital threats and harassment against employees in the police and child welfare, but to a lesser extent to threats against other private individuals and other professions than the aforementioned.

*Hate speech*, as a persecutory, hateful or degrading speech directed towards certain group attributes or "differences", such as race, gender etc., poses a real threat to freedom of speech. And it is on hate speech, especially hate speech *online*, that the spotlight has been directed at, both in the media and in the Norwegian courts last year. A newly published survey by the Institute of Social research shows that the issue of hate speech might be less acute than it may seem, and that it has perhaps been unduly amplified in the media.[82] But the same report is not totally representative of the situation either, since it is specified that young adults, low educated and people with immigrant background are under-represented in the survey. Moreover, another 2021 survey shows that about 5 % of the Norwegian population regularly participates in online debates with a harsh tone. Between 1 and 2 % of the respondents declare that they themselves

80 On July 22, 2011, several terrorist attacks were performed by a right-wing extremist, Anders Behring Breivik, that led to the death of 77 young people, mostly from the AUF, the youth division of the then ruling Norwegian Labour Party (AP), and to more than 200 persons injured.

81 One out of four journalists received threats in a land where free press is highly valued. See Mona Myklebust, Norske journalister trues med øks, pistol og avkuttet sauehode, NRK, 19.10.19, https://www.nrk.no/dokumentar/xl/norske-journalister-trues-med-oks_-pistol-og-avkuttet-sauehode-1.14729649 (last accessed: 26.05.21).

82 Audun Fladmoe, Dag Wollebæk, Kari Steen-Johnsen, Kjersti Thorbjørnsrud, Hva mener befolkningen

om ytringsfrihetens grenser?, Institutt for samfunnsforskning (ISF), Fritt Ords monitorprosjekt: Status for ytringsfriheten i Norge, April 2021, https://samfunnsforskning.brage.unit.no/samfunnsforskning-xmlui/bitstream/handle/11250/2738213/Hva%2Bmener%2Bbefolkningen%2Bom%2Bytringsfrihetens%2Bgrenser.pdf?sequence=1&isAllowed=y

engage in smear campaign online (*netthets*), in the form of provocative, racist or hate speech.[83]

The paradox is that, either way, both threats, hate speech, and their penalisation, impact on freedom of speech. On the one hand, threats and hate speech may directly impact the targets' equality, dignity, and autonomy in society,[84] but it can also have a chilling effect on their freedom of speech. On the other hand, the penalisation of hate speech and its enforcement by the courts may also have a chilling effect on those (either the political candidates, the elected politicians, or even ordinary people) who wish to express themselves on difficult, controversial issues. Another key-factor is that electoral campaigning calls for stronger or violent rhetoric and more polarised or exaggerated speech to convey a message more efficiently. Ordinary people in general, not just political debaters, may become hesitant to participate in the public debate and to hold more radical, controversial, or extreme opinions. Such a chilling effect is not in accordance with the idea of freedom of expression either.

### b. On the Current Legal Framework Regarding Digital Threats and Hate Speech

In practice, it is first and foremost the Norwegian Constitution and the ECHR that form the legal framework for freedom of expression in Norwegian law, with restrictions mainly stemming from the Penal code.

---

83 Marjan Nadim, Kjersti Thorbjørnsrud & Audun Fladmoe, Gråsoner og grenseoverskridelser på nettet: En studie av deltagere i opphetede og aggressive nettdebatter, Institutt for samfunnsforskning, Rapport 2021: 1, p. 7, https://samfunnsforskning.brage.unit.no/samfunnsforskning-xmlui/bitstream/handle/11250/2724997/Gr%25C3%25A5soner%2bog%2bgrenseoverskridelser%2bp%25C3%25A5%2bnettet.pdf?sequence=1&isAllowed=y; See Institute for Social Research, https://www.samfunnsforskning.no/aktuelt/arrangementer/2021/hvem-deltar-i-opphetede-nettdebatter.html; See also Audun Fladmoe, Dag Wollebæk,

Kari Steen-Johnsen & Kjersti Thorbjørnsrud, Hva mener befolkningen om ytringsfrihetens grenser, Institutt for samfunnsforskning, 2021, https://frittord.no/attachments/4c9c10582418489cff15e7a23105895f4a378956/192-20210416145734712684.pdf, https://frittord.no/nb/aktuelt/hvor-gar-ytringsfrihetens-grenser, https://www.mynewsdesk.com/no/institutt-for-samfunnsforskning/pressreleases/nye-tall-hvor-gaar-ytringsfrihetens-grenser-3091598

84 See Alon Harel, Hate Speech, in Adrienne Stone, Frederick Schauer (eds.), The Oxford Handbook on Freedom of Speech, *op. cit.*, Chapter 25, p. 456.

In Norway, like in many Western democracies, freedom of speech is strongly protected, diversity of opinion is highly valued and political utterances/statements (*frimodige ytringer*) enjoy a higher degree of constitutional protection than any other types of expression or speech.[85] Historically, the tolerance threshold for both racist and discriminatory speech has been very high, and there has been considerable reluctance to introduce too strict limitations on hate speech in Norway, as the regulation of hate speech implies limiting freedom of expression.

Freedom of expression is enshrined in article 100 of the Constitution. The *Convention of 4 November 1950 for the Protection of Human Rights and Fundamental Freedoms* (hereafter ECHR) and the *International Covenant of 16 December 1966 on Civil and Political Rights* (hereafter ICCPR), among others, have been incorporated in national law via the *Human Rights Act 1999*. The ECHR protects not only the right to freedom of expression but also the right to impart and receive ideas (art. 10 (1) ECHR), while the ICCPR protects both the right to hold opinions without interference and the right to freedom of expression, including "freedom to seek, receive and impart information and ideas of all kinds, regardless of frontiers, either orally, in writing or in print, in the form of art, or through any other media of his choice" (art. 19 ICCPR).

Yet, it follows from both these provisions and the case-law on the subject, that any interference with the freedom of expression, *i.e.*, a restriction thereof, must be authorized by law, pursue a legitimate purpose and be necessary. The condition of necessity means that any official limitation of free speech, e.g., in the form of punishment or penalty, must be both relevant and suitable for realizing its purpose, and proportionate. The right to freedom of expression may be restricted by a wide range of public interest considerations, including the "rights of others", see for example article 10 (2) ECHR and art. 19 (3) ICCPR. In addition, art. 20 (2) ICCPR declares that "[a]ny advocacy of national, racial or religious hatred that constitutes incitement to discrimination, hostility or violence shall be prohibited by law". Art. 17 ECHR prohibits the abuse of rights, and the ECtHR's case-law shows that there are cases where certain speech/utterances are either excluded from, or enjoy a restricted protection under, art. 10 ECHR.[86]

---

85 Article 100 (3) of the Constitution: «Everyone shall be free to speak their mind frankly on the administration of the State and on any other subject whatsoever. Clearly defined limitations to this right may only be imposed when particularly weighty considerations so justify in relation to the grounds for freedom of expression.» – Unofficial translation, https://lovdata.no/pro/#document/NLE/lov/1814-05-17

86 ECHR, Hate Speech, Factsheet, Sept. 2020, https://www.echr.coe.int/documents/fs_hate_speech_eng.pdf. See also art. 4 of CERD (UN Committee on the Elimination of

Freedom of expression is, to some extent, also limited by the Penal Code provisions on hate speech, threats, discriminatory statements, incitement to criminal acts, violations of privacy[87] and reckless conduct.

In Norwegian law, the prohibition of hate speech is mainly regulated in section 185 of the Penal Code.[88] It covers discriminatory or hateful expressions (hate speech) made in public or in the presence of others (including semi-public and private surroundings), in a grossly negligent or deliberate way. Such expressions, which can also take the form of pictures and symbols, may involve threatening or insulting someone or inciting hatred, persecution, or contempt of someone because of their skin colour or national or ethnic origin, religion or life stance, sexual orientation, gender identity or gender expression, or reduced functional capacity. The penalty may consist in a fine and/or imprisonment. Legal sanctions only apply for the most serious violations. In 2020, the Supreme Court of Norway issued no less than three rulings on hate speech, two of which on hateful utterances made on the social media.[89] The starting point is that it is mainly the person making the utterance that is responsible for its content (not intermediate actors such as the platforms or the moderators in the social media, for example). It is the number of persons who could have knowledge of the content of an utterance that is relevant, rather than where it was posted. According to section 10 subsection 2 PC: "If the act consists of making a statement, it is also public if the statement is made in a way that makes it likely to reach a sizable number of persons."[90] The definition of "public place" is still technology neutral in the law. Regarding the interpretation of the content of the statement,

---

Racial Discrimination), ratified by Norway and giving it a "clear obligation" to protect individuals and groups of people from discrimination based on ethnicity etc., through penal provisions such as the Penal Code's prohibition against hate speech (then Section 135a, now Section 185).

87 See footnote 71.

88 On the regulation of hate speech in Norway, see Jon Wessel-Aas, Audun Fladmoe and Marjan Nadim, *Hate Speech, Report 3. The boundary between freedom of speech and criminal law protection against hate speech*, Norwegian Institute for Social Research, Report 2016:22, p. 22 ff.

89 See HR-2020-184A and HR-2020-185A; Iris Nguyên Duy, The Limits to Free Speech on Social Media: On Two Recent Decisions of the Supreme Court of Norway, *Nordic Journal of Human Rights*, 2020, 38(3), pp. 237–245, https://doi.org/10.1080/18918131.2021.1872762

90 According to the preparatory works for the PC, when an utterance is published on the internet, it has been made 'publicly' if more than 20–30 people have had access to it. See Ot.prp nr 90 (2003–2004), section 12.2.2 and Prop53 L (2012–2013), point 3.6.

the Supreme Court's method relies on the average reader's or listener's natural perception of the statement made from the context.[91] The context in which the utterance is made is also crucial. Even when a statement qualifies as insulting, it is not punishable if it can be regarded as expressing a political statement or religious criticism, and, as such, taking part in the public debate. However, when an utterance is directed against the aggrieved party as a person, and the general reader would perceive it as highly degrading, it may fall under section 185 of the Penal Code. The Norwegian case law draws a line between critical statements on a subject, either of a political, cultural, religious nature or other, and statements attacking one or several persons. The first ones are normally protected by the freedom of expression and not covered by section 185 of the Penal Code, even if they are perceived as insulting.[92] The second type of utterances, depending on whether they fall under Section 185, enjoy only "modest constitutional protection", because they "do not have anything in common with the essence of what the freedom of expression is meant to protect, namely an open debate".[93]

The threshold for punishing hate speech has traditionally been high (even higher than in the ECHR's case law), and it is still high,[94] to avoid conflict with freedom of speech (hate speech must be aggravated or manifestly offensive), but it seems to have been lowered these past few years in the courts' practice. This can be explained by a normative evolution (§ 100 of the Constitution was revised in 2004, the Anti-Discrimination Act was adopted, and revised, a few years ago, a new criterium, gender identity and gender expression, was added to the exhaustive list in section 185 of the Penal Code last year) and by a more severe position of the courts in litigations on hate speech, especially concerning the penalty standard (a fine equivalent to one month's gross salary and/or some jail time) developed in recent Supreme Court case-law[95] – maybe influenced by a more systematic (and sometimes overdriven?) media cover of the issue.

---

91 In the Supreme Court's *Flyer* judgment (Rt. 1981 p. 1305, at p. 1314), the court refers to the "meaning the individual statements immediately convey to the not specially thoughtful reader", *i.e.*, the immediate, not significantly processed, understanding of an utterance.

92 See the Supreme Court judgment HR-2020-184-A § 24 and Rt-1981-1305, p. 1314.

93 See HR-2020-184-A § 25 and HR-2018-674-A, § 15 (the quarrel judgment).

94 Only statements of a "qualifiedly insulting nature" are covered by section 185 of the Penal Code, such as serious degradation of a group's human dignity or statements calling for or supporting integrity violations. But there is also a relatively wide margin for tasteless utterances. See the quarrel judgment, § 17 and the Supreme Court judgment Rt-2012-536 (the doorman judgment), § 28, § 29, § 32, § 33.

95 See HR-2020-184A, HR-2020-185A, HR-2020-2133-A.

The public discourse is quite divided on the penalisation of hate speech in Norway: Some advocate the suppression of this provision, because its definition is too vague, its scope, too large, and because it can have a chilling effect on those who wish to express themselves.[96] On the contrary, some others believe that a stricter penalization of hate speech would protect better the victims of hate speech.[97]

Other penal provisions are more or less directly applicable to hate speech and/or can be combined with section 185, such as section 263 of the Penal Code that stipulates that "[a]ny person who by words or conduct threatens to engage in criminal conduct under such circumstances that the threat is likely to cause serious fear shall be subject to a fine or imprisonment for a term not exceeding one year." In the case of aggravated threats, section 264 stipulates that anyone who threatens can be punished with imprisonment for up to 3 years. In deciding whether the threat is a serious (*grov*), special emphasis shall be placed on whether it is directed at a defenceless person, whether it is made unprovoked or by several people jointly, and whether it is motivated by the victim's skin colour, national or ethnic origin, religion or life stance, sexual orientation, gender identity or expression, or disability. The alternatives are not intended to be exhaustive; the decision is based on an overall assessment, and threats made based on someone's political affiliation may obviously also be relevant.

The Penal Code's general provisions on hate speech, threats, ruthless behaviour, or violation of privacy apply to all citizens.[98]

---

96 18 samfunnsdebattanter, Opprop: Rasismeparagrafen hører ikke hjemme i et liberalt demokrati, *Aftenposten*, 21.06.19, https://www.aftenposten.no/meninger/debatt/i/BRzx17/opprop-rasismeparagrafen-hoerer-ikke-hjemme-i-et-liberalt-demokrati

97 Note that "Research relating hate crime to the belief in a just world, however, clearly suggests that the absence of punishment may increase hate and hate crimes because it signals that the victim and even the whole group to which the victim belongs deserves this fate. This consequence is especially present for hate crimes because, in contrast with other crimes, the absence of punishment emphasizes the justification of the hatred." – Fischer, A., Halperin, E., Canetti, D., & Jasini, A. (2018). Why we hate, *op. cit.*, p. 136.

98 "Any person who by public communication violates the privacy of another person shall be subject to a fine or imprisonment for a term not exceeding one year.

The penalty pursuant to the first paragraph does not apply to a person who has participated only through technical assistance or distribution of a magazine or periodical produced within the realm. The same applies to broadcasts. " – See § 267 (1) and (2) of the Penal Code.

As for the position of the European Court on Human Rights, "there is a much greater willingness in the European jurisprudence to entertain a much broader range of justifications for restraints on speech, most of which are located in a sense of electoral fairness", observe professors Joo-Cheong Tham and KD Ewing.[99]

Liability for user-generated content is a relatively new area and the scope of the legislation in that respect has long been unclear.[100] There is no specific regulation sanctioning or penalizing the publication and/or dissemination of "fake news". However, according to the Gender Equality and Anti-Discrimination Ombud in its 2018-report on "Hate speech in the general debate online", when it comes to user-generated content that contains hate speech, editors with responsibility for the medium in which the content is published may be held liable, even when the editor does not have specific knowledge of the content.[101] Section 185 of the Penal Code on hate speech sets out, in addition to a requirement of intent, an alternative requirement of "gross negligence". "This means that an editor could be held criminally liable for user-generated content even when he does not have specific knowledge of the publication, if there was gross negligence".[102]

With the new, "technology neutral" (*i.e.*, regardless of the platform on which they are published on, see section 2 of the Act), *Media Liability Act 2020*,[103] the online debaters are responsible for their own statements, but the editors may have a certain co-responsibility and a certain "duty of care": an editor will not be held accountable if he removes or blocks access to, for example, hate speech,

---

99 Joo-Cheong Tham, KD Ewing, Free Speech and Elections, in Adrienne Stone, Frederick Schauer (eds.), *The Oxford Handbook of Freedom of Speech*, OUP, 2021, pp. 312–330, p. 330.

100 But regarding the editors' criminal liability, the most important provisions may be found in the recently adopted *Media Liability Act 2020* (in Norwegian: Lov 29. mai 2020 nr. 59 om redaksjonell uavhengighet og ansvar i redaktørstyrte journalistiske medier / Medieansvarsloven).

101 My translation. See Likestillings- og diskrimineringsombudet, Hatefulle ytringer i offentlig debatt på nett, *ldo.no*, 2018, p. 19, https://www.ldo.no/globalassets/_ldo_2019/arkiv/publikasjonsarkiv/fb-og-hatytring/ldo-fb_rapporten_2018.pdf (last accessed: 07.05.21).

102 *Ibidem*, p. 19.

103 LOV-2020-05-29-59 (entry into force: 1 July 2020), Act relating to the editorial independence and liability of editor-controlled journalistic media (The Media Liability Act). Unofficial translation available at: https://www.medietilsynet.no/globalassets/engelsk/210216-unoff-eng-versionact-relating-to-the-editorial-independence-and-liability-of-editor.pdf

speech that encourages violence and is a direct threat to journalists and others, in a comment field.[104] But if the editor chooses to ignore the content in a comment field, and leaves hateful remarks standing, he will, as an editor, have a shared responsibility. The editors have an *a posteriori* control function.

## *According to Section 6 (3) of the* Media Liability Act 2020

> The editor must facilitate the reporting of illegal user-generated content. If user-generated content is removed or access to the content is barred because it is considered illegal, the editor must, to the extent possible, notify the author of the content and inform him/her of opportunities to appeal.[105]

The whole chapter 4 of the Act contains rules on "Legal liability for content". For example, while section 12 (1) states that the media's technical contributors cannot be penalised for the "distribution of a statement only by way of technical preparation, production or distribution of the medium", section 13, designed on the model of the *E-Commerce Act 2003* [*ehandelsloven*] § 18, establishes criminal liability and liability for compensation for user-generated content:

> The editor, or any person acting on behalf of the editor, may only be penalised for illegal user-generated content in the medium if he or she has acted with intent. To be held liable for compensation, he or she must have acted with gross negligence.
> The editor, or any person acting on behalf of the editor, cannot be held liable pursuant to the first subsection if he or she without undue delay, once the conditions for liability were present, implements the necessary measures to remove or bar access to the illegal statement.

But only the editor-controlled media, *i.e.*, "media engaged in regular journalistic production and publication of news, current affairs, public debate or other content of general interest", are covered by this new Act (see section 2 of the Act), which means that the so-called "alternative media" are excluded if they do not fully meet the conditions of the law.

---

104 "According to the preparatory work, the provision will consequently not apply if the medium shares content on third-party platforms, such as Facebook, Twitter and YouTube. This means that if a media house shares an article, for example on Facebook, the obligation will not apply directly to statements made in Facebook's comment field."(my translation) – Hans Erik Johnsen, Kristiane Aarvik, Fride Støve Hedin, Ny medieansvarslov – moderne og mer effektiv, *Wiersholm*, 02.07.20, https://www.wiersholm.no/nyhetsbrev/ny-medieansvarslov-moderne-og-mer-effektiv

105 Unofficial translation: https://www.medietilsynet.no/globalassets/engelsk/210216-unoff-eng-versionact-relating-to-the-editorial-independence-and-liability-of-editor.pdf

Should the internet companies have a "duty of care"?[106] There is currently no specific regulation addressing the liability of the big media platforms in Norway for the digital services they provide as intermediaries and that emphasises explicitly and directly their responsibility for the content posted on their platforms in Norway either.[107] As a rule, the giants of the internet have been left to adopt their own policy regulations, and, in 2016, several internet giants, the IT companies Facebook, Microsoft, Twitter and YouTube, agreed with the European Commission on a code of conduct on countering illegal hate speech online, setting several public commitments.[108] The platforms agreed to assess the majority of user notifications in 24h (respecting EU and national legislation on hate speech) and committed to remove, if necessary, those messages assessed as being illegal.

But the situation is about to change, and it will probably go in the direction of more regulation, of *inter alia* Big Tech companies.

Norway has recently expressed its support of the European Union's initiative to develop a new regulatory framework for digital platforms, the *Digital Services Act*. But the Norwegian government has insisted that it should be built on the fundamental principles of the current *Electronic Commerce Directive*

---

106 See the British White Paper on, where the Government wishes to "establish a new statutory duty of care to make companies take more responsibility for the safety of their users and tackle harm caused by content or activity on their services." – Online Harms White Paper, 2020, https://www.gov.uk/government/consultations/online-harms-white-paper/online-harms-white-paper

107 Germany, France and Italy have adopted legislative measures that obliges social network providers to provide a system to handle complaints about illegal content: In Germany social media platforms that do not comply with the obligation to remove "manifestly unlawful" content within 24 hours long faced huge fines (see Network enforcement Act 2017 (NetzDG) section 3 (2) 2.). Such legislative measures arguably limit freedom of speech because of the danger of overreporting and of removing more content than necessary to avoid being fined, as the deletion of lawful content (such as satire) amounts to a form of censorship. The German legislation was amended in 2020. But see Amélie Heldt, Germany is amending its online speech act NetzDG… but not only that, *Internet Policy Review*, 6 April 2020, https://policyreview.info/articles/news/germany-amending-its-online-speech-act-netzdg-not-only/1464; Madeline Earp, Germany revisits influential internet law as amendment raises privacy implications, *CPJ*, 7 October 2020, https://cpj.org/2020/10/germany-revisits-influential-internet-law-as-amendment-raises-privacy-implications/

108 Currently, nine companies adhere to the Code: Facebook, YouTube, Twitter, Microsoft, Instagram, Google+, Dailymotion, Snapchat, and Jeuxvideo.com.

(entailing, *inter alia*, the continuation of the principles of limited liability for service providers when acting as intermediaries, and a requirement for the rules applicable in the country of origin to be followed), due to the importance of protecting free speech and democracy.[109] On the liability regime of digital services acting as intermediaries, the Norwegian government has underlined that

> It is important that any potential provisions on platforms' liability do not lead to platforms exercising editorial control over content made available by editorial media (such as takedown or modification). The DSA should establish safeguards which prevent platforms from applying additional or secondary control over such content. The Commission should take particular care not to overrule national sector-specific liability schemes and other national sector specific laws that apply to editorial media as content providers, even when editorial media content is published on platforms. The Norwegian Media Liability Act provides detailed rules for all editorial media on editorial independence, on criminal liability and liability for damages, which hold the author, the editor or the publisher responsible for the content. The Norwegian Media Liability Act applies to all editorial media, both self-published editorial content and editorial content published via online platforms such as Facebook and Snapchat. Editorial publishers also adhere to national criminal law and self-regulated ethical standards for the press.[110]

And in April 2021, the European parliament passed a new anti-terror internet regulation (TERREG) to address the dissemination of terrorist content online.[111] From next year onwards, hosting service providers will have to remove or disable access to flagged terrorist content in all member states within one hour of receiving a removal order from the competent authority. Content uploaded for educational, journalistic, artistic or research purposes, or used for awareness-raising purposes, will not be considered terrorist content under these new rules – but in practice, a lot will depend on the existence of actual safeguards and judicial control. Internet platforms will not have a general obligation to monitor or filter content. However, when competent national authorities have established

---

109 Norway has responded to the European Commission concerning the regulation of digital platforms, *Regjeringen.no*, 9.11.20, https://www.regjeringen.no/en/aktuelt/norway-has-responded-to-the-european-commission-concerning-the-regulation-of-digital-platforms/id2783237/ There is a link to Norway's position paper on the subject.

110 Norway's views on the consultation regarding the Digital Services Act (DSA), October 2020, p. 4–5, https://www.regjeringen.no/contentassets/120f9ac7eb814687a109fb4eacf9edbe/posisjonsnotat.pdf

111 Regulation (EU) 2021/784 of the European Parliament and of the Council of 29 April 2021 on addressing the dissemination of terrorist content online, http://data.europa.eu/eli/reg/2021/784/oj

a hosting service provider is exposed to terrorist content, the company will have to take specific measures to prevent its propagation. It will then be up to the service provider to decide what specific measures to take to prevent this from happening, and there will be no obligation to use automated tools.[112] Norway will probably incorporate it in its national law.

## 3. A Short Presentation of Selected COVID-19 Pandemic-related Challenges to Freedom of Speech Ahead of the 2021 Parliamentary Election

Because of the pandemic and its variants, and as long as a large proportion of the population has not yet been fully vaccinated,[113] many people stay at home, work from home and even, for some, live a significant part of their everyday life (shopping, socialising etc.) online. They are thus more permeable to the abovementioned digital threats.

This last section focuses on the specific challenges created by the Covid 19-pandemic in the context of the 2021 parliamentary election – and on the available material and legal solutions.

### a. Lack of "contingency plan" in the *Election Act 2002*

To vote and to be elected by universal and equal suffrage without discrimination is a fundamental right in national and international law and constitutes one of the cornerstones of modern democracies. Because of the Covid 19-pandemic, many

---

112 Press Release, New rules adopted for quick and smooth removal of terrorist content online, 28.04.21, https://www.europarl.europa.eu/news/en/press-room/20210422IPR02621/new-rules-adopted-for-quick-and-smooth-removal-of-terrorist-content-online (last accessed: 18.05.21). For a critic of the regulation, se Stefania Koskova, Europe's New Online Terrorist Content Regulation: A Gift to Budding Autocrats, Reporting Democracy, 05.05.21, https://balkaninsight.com/2021/05/05/europes-new-online-terrorist-content-regulation-a-gift-to-budding-autocrats/

113 On May 20, 2021, approximatively 1,6 M (out of 5.4 M Norwegians) persons had received the first dose of the Covid 19-vaccine and approximatively 7,50,000 persons, the second dose. See the statistics here: https://www.fhi.no/sv/vaksine/koronavaksinasjonsprogrammet/koronavaksinasjonsstatistikk/#norge-antall-vaksinerte Update: On 2 August 2021, the numbers are much higher: 3,59 M persons have received the first dose and 1,8 M are fully vaccinated in Norway.

voters worldwide have already been or can physically be barred from voting – or will even renounce voting.[114] As explained in the introduction, this can be seen as an encroachment on their freedom of expression. State authorities must ensure (facilitate) that all persons with voting rights have an opportunity to vote also during the covid-19 pandemic. Yet, rather surprisingly, neither the Norwegian Constitution nor the *Election Act 2002* contain any rules on the consequences an extraordinary event (such as a major accident, extreme weather, epidemics, or terrorist attacks) would have for the organisation of an election in Norway – or even indications on what type of measures that should be taken in such cases.[115] On the contrary, section 54 of the Constitution craves that "[e]lections shall be held every fourth year. They shall be concluded by the end of September." The absolute formulation of the constitutional provision can hardly be interpreted as allowing any exceptions, even in the case of a new outburst of the pandemic in September 2021.

Actual issues not dealt with by the current Norwegian legislation are, for example, whether the election can be postponed, whether the voting period can be extended, whether one can decide to hold a re-election or even to extend the election after the election has already taken place.[116] These issues have been examined by the Election Law Committee in 2020, but a reform of the *Election Act* will only take place after the 2021 elections, not before (the same goes for the *Security Act 2018*). The following reform proposals are relevant for this paper: The majority of the committee members (all except one member) proposes to introduce personal elections in parliamentary elections. Giving voters influence over which representatives are to be elected, in addition to which party is to get a representative, is considered an extension of democracy. The Election Law Committee proposes important changes to the appeals system in electoral litigation. Today, there is no independent legal review of election complaints. The committee proposes to turn the National Electoral Board into an independent appeals body, which would be appointed by the Storting. The majority of the committee also proposes to lower the voting age to 16 years in

114 See Erik Asplund, Global Overview of COVID-19: Impact on elections, *idea.int*, 24.05.21, https://www.idea.int/news-media/multimedia-reports/global-overview-covid-19-impact-elections (last accessed 24.05.21).

115 LOV-2002-06-28-57, Lov om valg til Stortinget, fylkesting og kommunestyrer (valgloven). Unofficial translation of the Act relating to parliamentary and local government elections (Election Act) available at: https://lovdata.no/dokument/NLE/lov/2002-06-28-57

116 See NOU 2020: 6, Frie og hemmelige valg, p. 318.

municipal and county council elections (but not in parliamentary elections). With regard to the counting of votes, the committee believes that it should be legislated that all votes must be counted in two independent ways in Norwegian elections. Finally, the committee unanimously proposes a contingency provision (*beredskapshjemmel*), which would allow for postponing or extending election day if something extraordinary happened that would prevent a significant part of the voters from voting. The committee further proposed that it should be possible to hold re-elections in such cases.[117] All these proposals touch upon, more or less directly, freedom of expression, but have not yet been examined by the legislator.[118]

## b. The Regulatory Answer to COVID-19 Challenges to Freedom of Expression

Although not optimal, the current regulatory framework is not that unsatisfactory, and helps dealing with the COVID-19 challenges in the context of the parliamentary election.

First, one must observe that the legal frame for the organisation of elections in Norway already provides mechanisms and procedures that other countries do not necessarily have.

Second, should there be an emergency situation this year close to election day, there are a few other legal solutions available, one being customary, and the other, temporary.

### *Existing Mechanisms and Procedures to Deal with the Pandemic-related Challenges*

Opening for advance voting has been one the measures taken to tackle Covid-19 in other countries. *Advance voting* was already legally possible in Norway

---

117 Valglovutvalget, Pressemelding fra valglovutvalget, https://www.valglovutvalget.no/pressemelding-fra-valglovutvalget/

118 In addition, the following challenge may be observed, but it falls out of the specific context of the 2021 parliamentary election, is not specific to Norway, and therefore falls out of the scope of this paper: the cooperation between the States and the online companies to fight COVID disinformation (but also hate speech and other abuses online), on the ground that "Lies cost lives in a global pandemic", may have negative effects on freedom of expression, when removal of content amounts to selective censorship. See, for example, https://www.theguardian.com/media/2020/may/02/youtube-deletes-coronavirus-conspiracy-theorist-david-ickes-channel; https://www.rollingstone.com/culture/culture-features/mewe-anti-vaxxers-conspiracy-theorists-822746/

before the outbreak of Covid-19 and considered a normal voting method in the country.[119]

On election day, the electors must vote in the municipality in which they are entered as resident in the population register on 30 June the same year as the election. All voters are eligible for early voting. One can vote early in any municipality in Norway. The normal time period for advance voting starts on 10 august (it is expected that many Norwegians will vote this way this year). One can even vote from 1 July if one cannot vote within the normal time period for early voting or on election day. If this is the case, the relevant municipality needs to be contacted to arrange a time to go and vote.

The state must ensure that people who live abroad may vote, despite border closure or the difficulty to travel abroad, and that people who are quarantined or in isolation may also vote. People living abroad may vote early from 1 July up to and including 1 September. But their vote must reach the municipality before 5:00 pm the day after election day on 10 September.[120]

The first two paragraphs in section 8-1 of the Election Act state:

(1) Electors who so wish may vote in advance inland from 10 August in the year of the election and abroad and on Svalbard and Jan Mayen from 1 July in the year of the election.
(2) Advance voting inland, including on Svalbard and Jan Mayen, must take place no later than the last Friday before Election Day, and advance voting abroad must take place no later than the penultimate Friday before Election Day. (…)

However, *online voting* will not be an option. Norway tested online voting during the 2011 local elections and 2013 Storting elections in a limited number of municipalities.[121] Regardless of the positive feedback, the parliament voted against further use of online voting in 2014: the turn out was not greater than before (probably because it is already very easy to vote in Norway) and, most importantly, there was a political disagreement related to the actual conduct of

119 So far (August 13), 105.094 voters have voted in the 2021 parliamentary elections. This means that 2.7 % of those entitled to vote in this year's election have already voted. https://www.valg.no/aktuelt/siste-nytt/sa-mange-har-stemt/ (last accessed: 13.08.21).
120 Valgdirektoratet, Frequently asked questions (FAQs), https://www.valg.no/en/old/election-in-norway/you-and-the-election/slik-stemmer-du/frequently-asked-questions-faqs/#wherevoteearly
121 Regjeringen, Internet Voting Trials, 29.08.19, https://www.regjeringen.no/no/tema/valg-og-demokrati/internet-voting-trials/id2666749/

the election, trumping over the fact that the electorate found the method easy and convenient.[122] In the end, the politician's trust in the popular voting system was considered more important than the voters' wishes.[123] Last year, the Election Law Committee discussed again the possibilities and consequences of online voting, held against four key aspects of elections: accessibility, security, secrecy and openness. The committee also considered what online voting would entail, both in the polling station and outside the polling station. Most of the committee members dismissed the idea of introducing online voting. The current security and technological advances are not good enough to implement such a technological solution and to guaranty secret elections, especially now in the face of the danger of international meddling from foreign powers. "The fact that votes are cast on paper, which are stored, checked and counted several times, is a basic element that makes the choice safer."[124] Voting should, as a rule, take place in a polling station, not from home. There are too many security issues, and it would not be possible to guaranty that a person is not pressured to vote a certain way, if they could do so from home.[125] It would also be problematic to let some groups of people vote electronically.[126] The idea of introducing e-voting this year was thus, temporarily at least, ruled out.

## *An additional, Temporary Legislative "fix"*

Both the *Election Act 2002* and the *Local Government Act 2018* have recently been temporarily amended to address the material challenges created by the pandemic. The temporary legislative changes will facilitate that all municipalities can carry out the election in a safe and good way, also where voters and election

---

122 Signe Bock Segaard et alia, Internettvalg – Hva gjør og mener velgerne, ISF rapport 2014:07, Oslo: Institutt for samfunnsforskning, 2014, p. 46; Bård Amundsen, No more online voting in Norway, *sciencenorway.no*, 12.09.19, https://sciencenorway.no/election-politics-technology/no-more-online-voting-in-norway/1562253

123 Regjeringen, Ikke flere forsøk med stemmegivning over internett, 23.06.14, https://www.regjeringen.no/no/aktuelt/Ikke-flere-forsok-med-stemmegivning-over-Internett-/id764300/

124 NOU 2020: 6, p. 61. It is also confirmed in annex 6 of the same document by Anne-Kari Valdal, Hermann S. Wiencke, Chris Dale, Svein Tuastad, Trine Holo, Willy Røed, Bjørg Sandal (Proactima AS), Sikkerheten i demokratiske prosesser i Norge, in NOU 2020: 6, pp. 478–512.

125 *Ibidem*, p. 61.

126 Ibidem, p. 62.

staff are ill or in quarantine. The main aim was to ensure that voters in isolation and quarantine can vote in elections.

It has thus been decided *inter alia* that:

- Voters who are sick, disabled or in isolation due to covid-19 can apply to vote from "where they are" in Norway (i.e., home), see *Election Act* section 8-3 b (advance voting), section 9-6 a.
- Polling stations can be located outside the geographical area of the constituency (Election Act section 9-3 a) and may consist of several rooms (*Election Act* section 8-3 a, 9-3 d).
- Voters who are in quarantine due to covid-19 will not be able to vote in ordinary polling stations but in separate premises, outside the polling station or at a mobile polling station (section 9-3 c (1)). If they cannot reach such a place, they may apply to vote from "where they are" (i.e., home), see *Election Act* section 8-3 c (advance voting), section 9-3 c (3).
- In the event of a local outbreak of infection, the municipality may decide to establish an extra polling station on election day if it were necessary to increase capacity and disperse voters (*Election Act* section 9-3 b (1)).
- In the event of a local outbreak of infection, the municipality may, on the last weekend before the election, decide to receive votes at voters' homes using only one polling station, but only if this is necessary for voters to be able to vote.
- Municipalities can start counting advance votes the day before election day,"[i]f this can happen without contravening the principle of secret ballot" (*Election Act* section 10-5 a).[127]

These temporary measures seem adequate to compensate for the absence of contingency plan in the law.

The outbreak of covid-19 has made it difficult for political parties and unregistered groups to obtain signatures on list proposals for the parliamentary elections in 2021. A temporary law was also adopted to make it possible for political parties that did not receive sufficient support in the previous parliamentary elections, and unregistered groups to collect signatures on list proposals electronically.[128]

127 Regjeringen, Regler for gjennomføringen av valget i 2021, 23.04.21, https://www.regjeringen.no/no/aktuelt/regler-for-gjennomforingen-av-valget-i-2021/id2845790/

128 https://www.stortinget.no/no/Saker-og-publikasjoner/Saker/Sak/?p=80205

*An Emergency Solution to be Found in Constitutional Customary Law?*

The abovementioned Election Law Committee interestingly assumes that, if the strict requirements for constitutional emergency were met, some legal basis might be found in constitutional (customary) law (*ulovfestet konstitusjonell nødrett*), to postpone or extend an election if the event of a significant number of voters being barred from voting or abstaining from voting as a result of extraordinary events.[129] The same goes for the organisation of an eventual re-election.

In a convincing report delivered to the Norwegian Government, professor Eirik Holmøyvik does not completely dismiss the possibility to ground possible deviations from the time limits set in § 54 of the Constitution in constitutional emergency, in case of a new outbreak of COVID-19 on the D-day or close to Election day, but that the threshold must be high.[130] The current electoral system is relatively satisfying and flexible enough (especially with the new, temporary, rules added to the *Election Act*) to help tackle such an issue.

## Concluding Remarks

*"Free political speech is a prerequisite for democracy".*[131]
*"How can you have an opinion if you are not informed?"*
*(Hannah Arendt)*[132]

This paper has focused on the specific issue of the non-legitimate threats to freedom of expression, mainly understood as the ability to form an informed opinion and to express oneself or to vote accordingly, in the context of the 2021 national elections in a time of pandemic, in Norway.

The combined impact of the COVID-19 pandemic and of the exponential development of new technologies on the country, the individuals, and their

129 NOU 2020: 6, p. 319–320; p. 356–8.

130 Eirik Holmøyvik, Utgreiing om høvet til å utsetje eller forlenge valtinget eller å halde omval ved stortingsvalet i 2021 grunna koronasituasjonen, 03.12.20, p. 21, https://www.regjeringen.no/contentassets/29640a7233e04651b25ec426f2143eb0/utgreiing-om-konstitusjonell-naudrett-i-samband-med-stortingsvalet-i-2021-eirik-holmoyvik.pdf

131 Alon Harel, Hate Speech, in Adrienne Stone, Frederick Schauer (eds.), *The Oxford Handbook on Freedom of Speech*, *op. cit.*, Chapter 25, p. 457–8.

132 On freedom of the press, but also applies to freedom of expression. Hannah Arendt from an Interview with French writer Roger Errera in 1974, *New York Review of Books*, 26.10.1978, https://www.nybooks.com/articles/1978/10/26/hannah-arendt-from-an-interview/

fundamental rights, has contributed to unveil or even exacerbate some of the fragilities of the current electoral system and legislation – and to renew interest in existing issues relating to freedom of expression.

On the one hand, technological developments have undoubtedly enhanced freedom of expression online. On the other, there are inequalities in access to and use of technology, even in a well-developed country such as Norway. Changing technology has created new problems associated *inter alia* with geography (no borders online) and scale of dissemination of information. The manipulation of both people and information has become much easier. All these developments have had an impact on freedom of speech. And the legislation in Norway only partially addresses the threats they sometimes represent. So far, the Norwegian response to the contemporary threats to freedom of expression has been rather "low-key" (but more intense in the case of hate speech) to avoid a chilling effect. This shows that freedom of expression is highly valued in Norway, that the authorities have not (yet) fallen for a "quick fix solution", but also that, in some areas, some reforms may be (are) necessary.

A reform of the electoral process is on its way, and Norway will probably incorporate the coming EU-legislation on e.g., the regulation of the digital platforms. The government has also recently mandated a commission to conduct a comprehensive review of the conditions for freedom of the speech.[133] Its members are asked to consider measures to promote the broadest possible participation in the public debate and means to hamper the spread of fake news and hate speech. They are specifically requested to *inter alia* investigate how to safeguard journalists' safety, as a precondition for medias watchdog function, and to what extent threats and incitement can lead to self-censorship.[134] But that will be the topic for a future article.

133 A commission on academic freedom has also newly been nominated. See https://www.regjeringen.no/no/aktuelt/anine-kierulf-skal-lede-ekspertgruppen-som-skal-se-pa-akademisk-ytringsfrihet/id2866450/ (last accessed: 02.08.2021).

134 The mandate of the commission may be found here: https://www.regjeringen.no/no/tema/kultur-idrett-og-frivillighet/innsiktsartikler/ytringsfrihetskommisjonen/mandatet/id2695703/ (last accessed: 26.05.21). Although Norway has been ranked nr. 1 in the Worldwide Press Freedom Index in 2021 (https://rsf.org/en/ranking), the problematic relationship between the press and the Norwegian government during the pandemic, as well as the lack of access to information on crisis management restrictions, could also be the topic of an article on the journalists and freedom of expression.

www.ingramcontent.com/pod-product-compliance
Ingram Content Group UK Ltd.
Pitfield, Milton Keynes, MK11 3LW, UK
UKHW022059051026
15623UKWH00003B/574

9 783631 870945